THE CRAFT OF INTELLIGENCE

Other Works by Allen W. Dulles

The Boer War: A History (1902)
Can We Be Neutral? (1936)
(with Hamilton Fish Armstrong)
Germany's Underground (1947)
The United Nations (1947)
Challenge of Soviet Power (1959)
The Secret Surrender (1966)
Great True Spy Stories (1968)
Great Spy Stories from Fiction (1969)

THE CRAFT OF INTELLIGENCE

AMERICA'S LEGENDARY SPY MASTER ON THE FUNDAMENTALS
OF INTELLIGENCE GATHERING FOR A FREE WORLD

ALLEN W. DULLES

Guilford, Connecticut

*To the men and women of the Central Intelligence Agency who are
devoting their careers to the building of American Intelligence*

An imprint of Rowman & Littlefield

Distributed by NATIONAL BOOK NETWORK

British Library Cataloguing in Publication Information Available

Library of Congress Cataloging-in-Publication Data Available

ISBN 978-1-4930-1879-6 (paperback)

∞™ The paper used in this publication meets the minimum requirements of
American National Standard for Information Sciences—Permanence of
Paper for Printed Library Materials, ANSI/NISO Z39.48-1992.

Contents

Preface: A Personal Note

My interest in world affairs started early; in fact, it goes back to my childhood days. I was brought up on the stories of my paternal grandfather's voyage of 131 days in a sailing vessel from Boston to Madras, India, where he was a missionary. He was almost shipwrecked on the way. In my youth, I was often in Washington with my maternal grandparents. My grandfather, John W. Foster, had been Secretary of State in 1892 under President Harrison. After serving in the Civil War he had become a general and had later been American minister to Mexico, to Russia and then to Spain. My mother had spent much of her youth in the capitals of these countries, my father had studied abroad. I grew up in the atmosphere of family debates on what was going on in the world.

My earliest recollections are of the Spanish and Boer Wars. In 1901, at the age of eight, I was an avid listener as my grandfather and his son-in-law, Robert Lansing, who was to become Secretary of State under President Woodrow Wilson, hotly discussed the merits of the British and Boer causes. I wrote out my own views—vigorous and misspelled—which were discovered by my elders and published as a little booklet; it became a "best seller" in the Washington area. I was for the "underdog."

After graduating from college a few months before the outbreak of World War I in 1914, sharing the general ignorance about the dramatic events that lay ahead, I worked my way around the world, teaching school in India and then China, and traveling widely in the Far East. I returned to the United States in 1915; and a year before our entry into the war, I became a member of the diplomatic service.

During the next ten years I served in a series of fascinating posts: first in Austria-Hungary, where in 1916–17 I saw the beginnings of the breakup of the Hapsburg monarchy; then in Switzerland during the war days, I gathered intelligence on what was going on behind the fighting front in Germany, Austria-Hungary and the Balkans. I was, in fact, an intelligence officer rather than a diplomat. Assigned to the Paris Peace Conference in 1919 for the Versailles Treaty negotiations, I helped draw the frontiers of the new Czechoslovakia, worked on the problems created for the west by the Bolshevik revolution of 1917 and helped on the peace settlement in Central Europe. When the Conference closed, I was one of those who opened our first postwar mission in Berlin in 1920, and after a tour of duty at Constantinople I served four years as Chief of the Near East Division of the State Department.

By that time, 1926, although I had still not exhausted my curiosity about the world, I had exhausted my exchequer and turned to the practice of the law with the New York law firm of which my brother was the senior partner. This practice was interrupted for periods of government service in the late twenties and early thirties as legal adviser to our delegations at the League of Nations conferences on arms limitations. In connection with this work I met Hitler, Mussolini, Litvinov and the leaders of Britain and France.

It was not only in the practice of the law that I was closely associated with my brother, John Foster Dulles. Though he was five years older than I, we spent much of our youth together. During the summers in the early 1900s and thereafter, as work permitted, Foster and I were together at the family's rustic summer quarters at Henderson Harbor on the southeastern shore of Lake Ontario. John W. Foster had started the Henderson Harbor family retreat before the turn of the century, in part because of his passion for smallmouth bass fishing, a trait which my brother and I inherited. Soon he was joined there by my father and mother and their five children of whom my brother, Foster, was the eldest. Mr. Foster's son-in-law, Robert Lansing, and my aunt, Mrs. Eleanor Foster Lansing, completed the contingent of the elder generation.

Here in delightful surroundings we indulged ourselves not only in fishing, sailing and tennis, but in never-ending discussions on the great world issues which our country was then growing up to face. These discussions were naturally given a certain weight and authority by the voices of a former Secretary of State and, after 1915, a Secretary of State in office. We children were at first the listeners and the learners, but as we grew up we became vigorous participants in the international debates. My brother, Foster, was often the spokesman for the younger generation on these occasions.

We were together in Paris in 1908–09 when Foster was doing graduate work in the Sorbonne and I was preparing for Princeton at the Ecole Alsacienne. From 1914 to 1919 our paths separated as I traveled around the world and later joined my diplomatic post in Vienna. But we had a reunion at the Paris Peace Conference in 1919. Our tasks there were different. He worked on the economic and financial issues of the peace and I largely on the political and new boundary questions. This association was precious to me and continued through the ensuing years. We later served together when in 1953 he became President Eisenhower's Secretary of State, and I was promoted from my job of Deputy, in which I had served under President Truman, to that of Director of Central Intelligence.

Deeply concerned with the basic issues of our times, with the tragedy of two fratricidal wars among the most highly developed countries of the world, Foster early saw grave new dangers to peace in the philosophy and policies of Communism. He became a convinced supporter of the work of the new Central Intelligence Agency. He wanted to check his own impressions and those of his associates in the State Department against an outside factual analysis of the problems which the President and he were facing. As a highly trained lawyer, he was always anxious to see the strength of all sides of an argument. He did not carry a foreign policy around in his hat. He sought the testing of his views against the hard realities of intelligence appraisals which marshaled the elements of each crisis situation. It was the duty of intelligence to furnish just this to the President and the Secretary of State.

Both Foster and I, in the course of our earlier years in law, diplomacy and international work, had been deeply influenced by the principles of Woodrow Wilson. We were thrilled with the high purpose he took to the Paris peace negotiations, where his first and main objective was the creation of the League of Nations to police a peace. We shared the frustrations of the Versailles negotiations, which, despite everything President Wilson could do, failed to provide a real basis for peace. My brother had fought, as had his colleagues on the Peace Delegation, against the unrealistic reparations clause of the treaty. At this time I was working on what seemed to me almost equally unsatisfactory territorial decisions, as the victors imposed the boundaries of the Versailles Treaty. All of this, as we could then only vaguely see, did much toward building up the bitterness that brought a Hitler to power and war to Europe in 1939.

When war threatened us in 1941, President Franklin D. Roosevelt summoned Colonel (later Major General) William J. Donovan to Washington to develop a comprehensive intelligence service. As the organizer and director of the Office of Strategic Services during World War II, Bill Donovan, I feel, is rightly regarded as the father of modern United States intelligence. After Pearl Harbor he asked me to join him, and I served with him in the OSS until the wars against Germany and Japan were over.

During these four demanding years I worked chiefly in Switzerland and after the German armistice in Berlin. I believe in the case history method of learning a profession, and here I had case after case, and I shall make use of them to illustrate various points in this narrative. Following the armistice with Japan, I returned to New York and the practice of law. This, however, did not prevent me from playing an active role in connection with the formulation of the legislation setting up the Central Intelligence Agency in 1947.

The following year, President Truman asked me to head up a committee of three, the other two members being William H. Jackson, who had served in wartime military intelligence, and Mathias F. Correa, who had been a special assistant to the Secretary of the Navy, James Forrestal. We were asked to report on the effectiveness of the CIA as organized under the 1947 Act and the relationship of CIA activities to those of other intelligence organs of the government.

Our report was submitted to President Truman upon his reelection and I returned once again to full-time practice of the law, expecting this time to stay with it. But writing reports for the government sometimes has unexpected consequences. You may be asked to help put your recommendations into effect. That is what happened to me. Our report suggested some rather drastic changes in the organization of CIA, particularly in the intelligence estimative process. General Walter Bedell Smith, who had become Director in 1950, and already had appointed Jackson as his deputy, invited me down to discuss the report with him. I went to Washington intending to stay six weeks. I remained with CIA for eleven years, almost nine years as its Director.

Since returning to private life in November of 1961, I have felt that it was high time that someone—even though he be a deeply concerned advocate—should tell what properly can be told about intelligence as a vital element of the structure of our government in this modern age.

In writing this book as a private citizen I wish it to be clearly understood that the views expressed are solely my own and have not been either authorized or approved by the Central Intelligence Agency or any other government authority.

This revised edition of *The Craft of Intelligence*, prepared over a year after the first edition went to press in 1963, contains a considerable amount of new material. In some instances, in the interim, events and issues I described earlier—for example, the swapping of captured spies—had developed in such a fashion that it would be a serious omission not to bring them up to date; in other instances, cases which had not been publicly disclosed were surfaced in the press as accused spies came to trial, and I was now free to speak of them.

1
The Historical Setting

In the fifth century B.C. the Chinese sage Sun Tzu wrote that foreknowledge was "the reason the enlightened prince and the wise general conquer the enemy whenever they move." In 1955, the task force on Intelligence Activities of the second Herbert Hoover Commission in its advisory report to the government stated that "Intelligence deals with all the things which should be known in advance of initiating a course of action." Both statements, widely separated as they are in time, have in common the emphasis on the practical use of advance information in its relation to action.

The desire for advance information is no doubt rooted in the instinct for survival. The ruler asks himself: What will happen next? How will my affairs prosper? What course of action should I take? How strong are my enemies and what are they planning against me? From the beginnings of recorded history we note that such inquires are made not solely about the situation and prospects of the single individual but about those of the group—the tribe, the kingdom, the nation.

The earliest sources of intelligence, in the age of a belief in supernatural intervention in the affairs of men, were prophets, seers, oracles, soothsayers and astrologers. Since the gods knew what was going to happen ahead of time, having to some extent ordained the outcome of events, it was logical to seek out the divine intention in the inspiration of holy men, in the riddles of oracles, in the stars and often in dreams.

Mythology and the history of religion contain countless instances of the revelation of the divine intention regarding man, solicited or unsolicited by men themselves. But not many of them have to do with the practical affairs of state, with the outcome of military ventures and the like. Yet there are some, and I look upon them as the earliest recorded instances of "intelligence-gathering."

Saul, on the eve of his last battle, "was afraid, and his heart greatly trembled" when he saw the host of the Philistines. "And when Saul enquired of the Lord, the Lord answered him not, neither by dreams, nor by Urim, nor by prophets" (I Sam. 28) Being without "sources" and wondering what course to follow in the battle to come, Saul, as we all know, summoned up the spirit of Samuel through the witch of En-dor and learned from him that he would lose the battle and would himself perish. In a subsequent chapter of the Book of Samuel we find David directly questioning the Lord for military advice and getting exactly the intelligence he needed. "Shall I pursue after this troop? Shall I overtake them? And he [the Lord] answered him, Pursue: for thou shalt surely overtake them, and without fail recover all."

An even earlier "intelligence operation" recorded in the Bible is of quite another sort (Num. 13). Here the Lord suggested that man himself seek information on the spot.

When Moses was in the "wilderness" with the children of Israel, he was directed by the Lord to send a ruler of each of the tribes of Israel "to spy out the land of Canaan," which the Lord had designated as their home. Moses gave them instructions to "see the land, what it is; and the people that dwelleth therein, whether they be strong or weak, few or many." They spent forty days on their mission. When they came back; they reported on the land to Moses and Aaron: "Surely it floweth with milk and honey; and this is the fruit of it"—the grapes, the pomegranates and the figs. But then ten of the twelve who had gone on this intelligence mission, with Joshua and Caleb dissenting, reported that the people there were stronger than the men of Israel.

They were "men of great stature," and "the cities are walled and very great," and "the children of Israel murmured against Moses and against Aaron." The Lord then decreed that because of the little faith that the people had shown in him they should "wander in the wilderness forty years," one year for every day that the spies had searched the land, only to bring in their timorous findings.

In this particular intelligence mission, there is more than meets the eye at first reading. To begin with, if one wanted a fair and impartial view of the nature of the land of Canaan and its people, one would not send political leaders on an intelligence mission. One would send technicians, and surely not twelve, but two or three. Furthermore, Moses and Aaron did not need information about the land of Canaan, as they trusted the Lord. The real purpose of this mission was, in fact, not to find out what sort of a land it was: it was to find out what sort of people—how strong and trustworthy—were these leaders of the various tribes of Israel. When only two met the test in the eyes of the Lord, the rest and their peoples were condemned to wander in the desert until a new and stronger generation arose to take over.

It is a part of history that intelligence even when clear should all too often be disregarded or sometimes not even sought. Cassandra, the daughter of Priam of Troy, who was beloved by Apollo, was accorded by him the gift of prophecy. But, as mythology tells us, once she had obtained the gift, she taunted the tempter. Apollo could not withdraw his gift but could and did add to it the qualification that her prophecies should not be believed. Hence, Cassandra's prediction that the rape of Helen would spell the ruin of Troy and her warning about the famous Trojan Horse—one of the first recorded "deception" operations—were disregarded.

The Greeks, with their rather pessimistic view of man's relations with the gods, seem to have run into trouble even when they had information from the gods because it was so wrapped in riddles and contradictions that it was either ambiguous or unintelligible. The stories about "intelligence" that run through Greek mythology reflect a basic conviction that the ways of the gods and of fate are not for man to know.

Herodotus tells us that when the Lacedaemonians consulted the Delphic oracle to learn what the outcome of a military campaign against Arcadia would be, the oracle answered that they would dance in Tegea (a part of Arcadia) with "noisy footfall." The Lacedaemonians interpreted this to mean that they would celebrate their victory there with a dance. They invaded Tegea, carrying fetters with which to enslave the Tegeans. They lost the battle, however, and were themselves enslaved and put to work in the fields wearing the very fetters they had brought with them. These, shackled about their feet and rattling as they worked, produced the "noisy footfall" to which the oracle had referred.

Over the centuries the Delphic oracle evolved through a number of stages, from a "supernatural" phenomenon to an institution that was apparently more human and more secular. In its earliest days a virgin sitting over a cleft in the rock from which arose intoxicating fumes received in a trance the answers of the god Apollo to the questions that had been asked, and a priest interpreted the magical and mysterious words of the "medium." The possibility of error and prejudice entering at this point must have been great. Later the virgins were replaced by women over fifty because the visitors to the oracle seem to have disturbed its smooth operation by an undue and strongly human interest in the virgins. But that did not necessarily affect the allegedly divine nature of the revelations given. What did make the oracle more of a secular institution at a later date, as we know today, was the fact that the priests apparently had networks of informants in all the Greek lands and were thus often better appraised of the state of things on earth than the people who came for consultation. Their intelligence was by no means of divine origin, although it was proffered as such. At a still later stage, a certain corruption seems to have set in as a result of the possession on the part of the priests of the secrets which visitors had confided to them. A prince or a wealthy man who either was favored by the priests at Delphi or perhaps bribed them could have picked up information about his rivals and enemies which the latter had divulged when they consulted the oracle. In their most productive period, the oracles frequently produced excellent practical advice.

But in the craft of intelligence the East was ahead of the West in 400 B.C. Rejecting the oracles and the seers, who may well have played an important role in still earlier epochs of Chinese history, Sun Tzu takes a more practical view.[1]

"What is called 'foreknowledge' cannot be elicited from spirits, nor from gods, nor by analogy with past events, nor from calculations," he wrote. "It must be obtained from men who know the enemy situation."

In a chapter of the *Art of War* called the "Employment of Secret Agents," Sun Tzu gives the basics of espionage as it was practiced in 400 B.C. by the Chinese—much as it is practiced today. He says there are five kinds of agents: native, inside, double, expendable and living. "Native" and "inside" agents are similar to what we shall later call "agents in place." "Double," a term still used today, is an enemy agent who has been captured, turned around and sent back where he came from as an agent of his captors. "Expendable agents" are a Chinese subtlety which we later touch upon in considering deception techniques. They are agents through whom false information is leaked to the enemy. To Sun

[1]For my remarks on Sun Tzu I am indebted to the recent excellent translation of the *Art of War* with commentaries by General Sam Griffith (Oxford, Clarendon Press, 1963).

Tzu they are expendable because the enemy will probably kill them when he finds out their information was faulty. "Living" agents to Sun Tzu are latter-day "penetration agents." They reach the enemy, get information and manage to get back alive.

To Sun Tzu belongs the credit not only for this first remarkable analysis of the ways of espionage but also for the first written recommendations regarding an organized intelligence service. He points out that the master of intelligence will employ all five kinds of agents simultaneously; he calls this the "Divine Skein." The analogy is to a fish net consisting of many strands all joined to a single cord. And this by no means exhausts Sun Tzu's contribution. He comments on counterintelligence, on psychological warfare, on deception, on security, on fabricators, in short, on the whole craft of intelligence. It is no wonder that Sun Tzu's book is a favorite of Mao Tse-tung and is required reading for Chinese Communist tacticians. In their conduct of military campaigns and of intelligence collection, they clearly put into practice the teachings of Sun Tzu.

Espionage of the sort recommended by Sun Tzu, which did not depend upon spirits or gods, was, of course, practiced in the West in ancient times also, but not with the same degree of sophistication as in the East; nor was there in the West the same sense of a craft or code of rules so that one generation could build on the experiences of another. Most recorded instances do not go far beyond what we would call reconnaissance. Such was the case in the second and more successful attempt of the Israelites to reconnoiter the situation in the Promised Land.

Joshua sent two men into Jericho to "spy secretly," and they were received in the house of Rahab the harlot (Josh. 2). This is, I believe, the first instance on record of what is now called in the intelligence trade a "safe house." Rahab concealed the spies and got them safely out of the city with their intelligence. The Israelites conquered Jericho "and utterly destroyed it and its people except that Rahab and her family were saved." Thus was established the tradition that those who help the intelligence process should be recompensed.

According to Herodotus, the Greeks sent three spies to Persia before the great invasion of 480 B.C. to see how large the forces were that Xerxes was gathering. The three spies were caught in the act and were about to be executed when Xerxes stayed their execution and to the great surprise of his counselors had the spies conducted all around his camp, showing them "all the footmen and all the horse, letting them gaze at everything to their hearts' content." Then he sent them home. Xerxes' idea was to frighten the Greeks into surrendering without a fight by deliberately passing them correct information as to the size of the host he had assembled. Since, as we know, the Greeks were not intimidated, he did not succeed in this psychological ploy. I have an idea that Sun Tzu would have advised the opposite. He would have recommended that Xerxes bribe the spies and send them home to report that this army was far smaller and weaker than it really was. When the Persians later invaded, Sun Tzu would have expected the three men to report to him what was going on in the Greek camp.

Just before the battle of Thermopylae, Xerxes himself sent a "mounted spy" to see what the Greeks, who were holding the pass, were doing and how strong they were. This was clearly nothing but a short-range reconnaissance mission. But Xerxes' scout got very close because when he returned he was able to give the famous report that some of the men he saw were "engaged in gymnastic exercises, others were combing their long hair." This was a piece of "raw intelligence," as we would call it today, that obviously stood in need of interpretation and analysis. Accordingly, Xerxes called in one of his advisers who knew Greek ways and who explained to him that "These men have come to dispute the pass with us; and it is for this that they are now making ready. It is their custom, when they are about to hazard their lives, to adorn their heads with care. . . . You have now to deal with the first kingdom in Greece, and with the bravest men." Xerxes did not put much faith in the "estimate" and lost vast numbers of his best troops by throwing them directly against the little band of Greeks under Leonidas.

Altogether in the Western world in ancient times the use and the extent of espionage seems to have depended on the personality and strength and ambition of kings and conquerors, on their own propensity for wiles and stratagems, their desire for power and the need to secure their kingdoms. Athens in the days of democracy and Rome in the days of the republic were not climates that bred espionage. Government was conducted openly, policy made openly, and wars usually planned and mounted openly. Except for the size and placement of enemy forces at key moments before the engagement in battle there was little need felt for specific information, for the foreknowledge that could affect the outcome of great exploits. But for the great conquerors, the Alexanders and the Hannibals, the creators of upstart and usually short-lived empires, this was not so. Subject peoples had to be watched for signs of revolt.

Whirlwind campaigns which were frequently great gambles were more likely to succeed if one had advance knowledge of the strength and wealth of the "target" as well as the mood and morale of its rulers and populace. The evidence suggests that empire-builders such as Alexander the Great, Mithridates, King of Pontus, and Hannibal all used and relied to a much greater extent on intelligence than their predecessors and contemporaries. Hannibal, a master of strategy, is known to have collected information before his campaigns not only on the military posture of his enemies but on their economic condition, the statements in debate of public figures and even civilian morale. Time and again Plutarch makes mention of Hannibal's possession of "secret intelligence," of "spials he had sent into the enemies' camp."

Hannibal appears to have been weaker as a linguist than as a strategist. Plutarch tells us that while in Southern Italy Hannibal commanded his guides to take him to the plain of Casinum. (This was Cassino of World War II fame.) "They, mistaking his words . . . because his Italian tongue was but mean, took one thing for another and so brought him and his army . . . near the city of Casilinum." The terrain was such that Hannibal was nearly trapped, but he took time out to dispose of those who had misled him. "Knowing then the fault his guides had made and the danger wherein they had brought him, he roundly trussed them up and hung them by the necks." This story is often told today in intelligence schools to impress upon junior officers the need for accuracy.

Mithridates fought the power of Rome to a standstill in Asia Minor in part because he had become an outstanding intelligence officer in his own right. Unlike Hannibal, he mastered twenty-two languages and dialects and knew the local tribes and their customs far better than did the Romans.

During the Middle Ages, due as much to the fragmented political situation as to the difficulties of transportation, supply and mobilization, it was impossible to attain strategic surprise in military campaigns. It took weeks, even months, to assemble an army, and even when the force had been collected, it could move only a few miles a day. Seaborne expeditions could move somewhat more unobtrusively, but the massing of ships was difficult to conceal. For example, in 1066 King Harold of England had all the essential intelligence long before William the Conqueror landed at Hastings. He had been in Normandy himself and had seen the Norman Army in action. He knew that William was planning an attack; he estimated the planned embarkation date and landing place with great accuracy; and, judging by the size of the force he concentrated, he made a very good guess about the number of William's troops. His defeat was not due to strategic intelligence deficiencies. He lost, rather, because his troops were battle-weary. He had just beaten the Danes in a smashing victory at Stanford Bridge. Also, they were exhausted after a long forced march.

The most serious political mistakes of Western Europe in the Middle Ages were made in relation to the East, due in large part to inadequate intelligence collection. European rulers consistently weakened Byzantium instead of supporting it as a bulwark against invasion. They failed to recognize both the dangers and the opportunities created by the Mongol drive to the west. They underestimated the Turkish threat during the period when the Ottomans were consolidating their power. Given their prejudices, they might have made the same mistakes even if they had had better intelligence support, but without it they had almost no chance of making correct decisions.

They were not very well informed about the Byzantine Empire and the Eastern Slavs; they knew even less of the Moslem world, and they were almost completely ignorant of anything that went on in Central and East Asia. Emperor Frederick II (1212–50) tried to keep up contacts with Moslem rulers (and was denounced as a heretic for his pains), and Louis IX of France (1226–70) sent emissaries to the Mongols. Marco Polo's famous book about China contained material that would have been useful for strategic intelligence, but no one looked at it in that light. Throughout most of the Middle Ages Italian merchants did obtain considerable information about the East; unfortunately, they seldom had a chance to pass it on to the people who determined Europe's Oriental policy. The popes disliked the merchants' willingness to trade with enemies of the faith, and kings had little contact with them.

In the fifteenth century the Italians made an important contribution to intelligence collection by establishing permanent embassies abroad. The envoys of Venice were especially adept at obtaining strategic intelligence. Most of their reports were of a very high quality, full of accurate observations and shrewd judgments. Not only did permanent embassies provide for this kind of observation, but they also provided bases from which to establish regular networks of espionage. By the sixteenth century, most European governments were following the example of the Italian city-states.

Because map making was an almost unknown art in earlier times, an important item of intelligence was information on local geography.

Knowledge of a river ford might allow an army to escape encirclement; discovery of a mountain path could show the way past a strong enemy position. Local inhabitants could usually be induced to give this kind of information, and Louis IX gave a large reward to a Bedouin who showed him where to cross a branch of the Nile, thereby enabling him to stage a surprise attack upon a Moslem army. Louis' son turned a strong defensive position in the Pyrenees by buying information about a little-used route through the mountains. Better known is the incident in the Crécy campaign when Edward III was nearly hemmed in by a large French Army. A shepherd showed him a ford across the Somme, and Edward not only escaped pursuit but also obtained such a strong defensive position that he was able to break the French Army when it finally attacked.

With the rise of nationalism and the religious struggles of the sixteenth and seventeenth centuries, the first real specialists in intelligence began to appear on the Western scene—ministers and secretaries of cabinet who devoted much of their careers to organizing the collection of secret information. Because of the frequency of internal dissension and civil strife in this era, we also see at the same time the beginning of a distinction between foreign intelligence and internal security. It was still too soon for the existence of two separate services with distinct responsibilities—that came later—but it was a period in which spies at home were as important as spies abroad, all of them manipulated by the same hand.

One of the masters of both arts was Sir Francis Walsingham, who spent most of his life as Secretary of State and chief spymaster in the service of Queen Elizabeth. Walsingham's hand can be discovered behind many of the major undertakings of Elizabeth's reign, preparing the ground, gathering the necessary information, provoking conspiracies and then exposing them. There is hardly a technique of espionage which cannot be found in his practice of the craft. Thanks to him the foolish and weakly conceived Babington conspiracy to bring Mary Queen of Scots to the English throne grew to such dimensions that it finally gave Elizabeth the pretext to sign Mary's death warrant. The most gifted graduates of Oxford and Cambridge were enlisted by Walsingham to study in France and to penetrate the French court and learn of its designs against England. Christopher Marlowe appears to have been one of them, and his premature death in a tavern brawl at Deptford is thought to have been the unfortunate result of one of Walsingham's plots.

Walsingham's greatest coup was undoubtedly the skillful roundabout operation which procured for England the naval intelligence on which its defense against the Spanish Armada was in great measure based. Instead of trying to strike directly against his target, the court of Philip II of Spain, Walsingham avoided the obvious, the direct reconnaissance tactic, so often doomed from the start, and operated through other areas where he knew there were vulnerabilities that could give him access to Spain. He dispatched a pair of young Englishmen to Italy who had excellent connections at the Tuscan court. (Throughout Walsingham's operations we find professed religious affiliations playing a major role, Protestants masquerading as Catholics and claiming to espouse the cause of England's enemies.) One of these young Englishmen, Anthony Standen, cultivated the Tuscan Ambassador to Spain with such success that he arranged for the employment of his agents with the latter's mission in Spain, thus infiltrating into the Spanish ports trustworthy observers who were not Englishmen and in no way would arouse suspicion of being in the service of the English. As a favor the Tuscan Ambassador even let Standen's "friends" in Spain use his diplomatic pouch to send "personal" letters to Standen in Italy.

Under Walsingham it became established practice for Her Majesty's Secretary of State to intercept domestic and foreign correspondence, to open it, read it, reseal it and send it on its way. Should such correspondence be in code or cipher, Walsingham had in his service an expert, a certain Thomas Phelippes, who was both cryptographer and cryptanalyst; that is, he invented secure codes for Walsingham's use and at the same time broke the codes used in messages which Walsingham intercepted. It was Phelippes who deciphered the rather amateurish secret messages which went to and from Mary Queen of Scots at the time of the Babington conspiracy.

Walsingham, in short, created the first full-fledged professional intelligence service. He was shortly after to be rivaled by Richelieu, but hardly by any other master of espionage until the nineteenth century.

Much has been made, to be sure, of Cromwell's intelligence chief, John Thurloe, but in the perspective of history I do not find him possessed of the same ingenuity, inventiveness and daring that distinguished Walsingham. A major key to Thurloe's success was the very sizable funds he had at his disposal. Pepys says he spent over £70,000 a year. This figure may be exaggerated, but the records show that he paid his spies inordinate sums for their information and thus had little difficulty recruiting them. Walsingham, on the other hand, worked with the most niggardly budget under the tight-pursed Queen and is said frequently to have paid his agents out of his own pocket, and then only insignificant sums.

Thurloe, like Walsingham, had the title of Secretary of State, but by this time his office had become known as the "Department of Intelligence," one of the earliest official uses of the designation in English for a bureau of government. His was, of course, a time of major conspiracies bent on restoring Charles Stuart to the throne. For this reason, again as in Washington's time, Thurloe ran both an internal security service and a foreign intelligence system. For the latter he used English consuls and diplomats abroad but supplemented their reporting with the work of secret agents. Thurloe relied even more than did Walsingham on information from postal censorship and can certainly be credited with having run a very efficient post office from the point of view of counterintelligence.

Despite the calm, almost humdrum way in which Thurloe seems to have gone about the business of systematic intelligence collection, he was frequently involved in heavy-handed plots. One of these, which he prepared at Cromwell's instigation, had as its purpose the assassination of Charles and the Dukes of York and Gloucester, his brothers. This was in reprisal for a Royalist plot directed against Cromwell's life which Thurloe had uncovered. The scheme was to entice the three royal brothers from France to England on the false claim that they would be met by a body of soldiers on landing who would then set off an uprising. It all sounds rather obvious and contrived at this distance and has none of the subtlety of Walsingham's plots in which he successfully involved Mary Queen of Scots. Whether Charles would have fallen for the trick we need not conjecture, because one of Thurloe's closest confidants, his secretary, Morland, betrayed the plot to Charles. Pepys tells us in his diary that only five days after Charles was restored to the throne, "Mr. Morland was knighted . . . and the King did give the reason of it openly, that it was for his giving him intelligence all the time he was clerk to Secretary Thurloe."

Another interesting example of successful seventeenth-century intelligence is that of Sweden, which maintained its position as a great power to a very considerable degree by virtue of having the most accurate reporting system in Europe. A contemporary Russian minister admitted that "the Swedes know more about us than we do ourselves." They played heavily on Protestant connections during the period of the religious wars and generally used men of other nationalities such as French Huguenots as both agents and reporters, much in the manner of Walsingham, thereby avoiding embarrassment and direct implication if caught. Sweden and to some extent Holland in those days illustrate how relatively small countries can make up for many power deficiencies with superior intelligence combined with technical and organizational ingenuity.

In the late eighteenth and early nineteenth centuries, an ever-sharpening distinction emerged between the work of internal security and the collection of foreign intelligence. In the major powers, separate organizations under separate experts were more and more entrusted with the different tasks. The reason, of course, was that the growth of internal dissidence, the threat of uprising and revolution from within, threatened the stability and power of the great autocratic and imperial systems of nineteenth-century Europe, thus causing the burgeoning of secret police organs for the protection of the emperor or ruler.

Under Napoleon, first the infamous Joseph Fouché, a product of the turbulent conspiracies of the French Revolution, and later Colonel Savary served as Ministers of Justice and chiefs of a purely political secret police and counterespionage organization. The collection of military and foreign intelligence, however, was in the hands of the Alsatian, Karl Schulmeister, who, though nominally attached to Savary, ran a quite autonomous series of operations whose purpose was to gain intelligence about the Austrian armies and to deceive the Austrians as to the strength and intentions of the French.

Gradually the growth of large and aggressive armed forces during the nineteenth century caused the emphasis in foreign intelligence to be placed primarily on its military aspects and the responsibility for its collection to be taken over by the army itself. In the period up to the outbreak of World War I, under the aegis of the General Staffs of most European armies a single military intelligence agency developed and became the major foreign intelligence arm of the country. It was directed by military officers rather than by civilians or cabinet ministers. Political intelligence was left largely to the diplomats.

Prussia up to 1871 was the exception to this development, primarily because the power-hungry, though gifted Wilhelm Stieber kept the reins of both Prussian military intelligence and of the Prussian secret police in his ambitious hands. To him goes the credit for the first exercises in mass espionage, for the method of saturating a target area with so many spies that they could hardly fail to procure detailed information on every aspect of an enemy's military and political status. These networks were also a kind of fifth column and helped soften the morale of civilian populations by inducing a fear of the coming invader. Previously, espionage had made use of a few selected and highly placed individuals. Stieber went after the farmers and the storekeepers, the waiters and the chamber-maids. He used these methods in preparing for the Prussian attacks against both Austria in 1866 and France in 1870.

The size and power of an internal security service is generally in direct ratio to the extent of the suspicion and fear of the ruling clique. Under a repressive and autocratic ruler secret police will blossom, a dreaded parasitical force that permeates every element of the populace and the national scene. For the best example of such an organization we must, therefore, turn to nineteenth-century Russia, where a retarded political system stood in constant fear of its own masses, its liberal leaders or the dangerous ideas and influences of its neighbors.

But this state of affairs in Russia was not an innovation of the nineteenth century. In early Russian history, the Tatars and other steppe people continually sought to ascertain the strength of the garrisons within the walled stockades (kremlins) of the Russians. As a result, the Russians became congenitally suspicious of anyone seeking admission to the walled cities, fearing that their real mission was intelligence. The tradition of attaching a *pristav* (literally, "an attached object") to a visiting foreigner, so that he could be readily identified as such, goes back at least to the sixteenth century. There is a long ancestry for surveillance and "guided tours" in Russia. In the seventeenth century, when the Russians began sending their own people abroad to study at foreign universities, they usually sent some trusted person along to watch and report on any group of students. The custom of attaching a secret policeman to delegations attending international conferences, so much in evidence today, therefore also has hoary antecedents.

An organized political police under state management in Russia can be traced back to the establishment in 1826 by Czar Nicholas I of the Third Section of His Majesty's Imperial Chancery. In 1878 the Third Section was abolished and its functions were given to the Okhrana, or security section, of the Ministry of the Interior.

The purpose of the Czar's Okhrana was to "protect" the imperial family and its regime. In this capacity it kept watch on the Russian populace by means of armies of informants, and once even distinguished itself by tailing the venerable Leo Tolstoi around Russia. Tolstoi had long since become a world-renowned literary figure, but to the Okhrana he was only a retired army lieutenant and a "suspect."

In the late nineteenth century there were so many Russian revolutionaries, radical students and *émigrés* outside Russia that the Okhrana could not hope to keep Imperial Russia secure merely by suppressing the voices of revolution at home. It had to cope with dangerous voices from abroad. It sent agents to join, penetrate and provoke the organizations of Russian students and revolutionaries in Western Europe, to incite, demoralize, steal documents and discover the channels by which illegal literature was being smuggled into Russia. When Lenin was in Prague in 1912, he unknowingly harbored an Okhrana agent in his household.

When Bolsheviks swept into power in 1917, they disbanded and to some extent "exposed" the old Okhrana as a typical oppressive instrument of the czars, claiming that the new workers, state needed no such sinister device to maintain law and order. In the same breath, however, they created their own secret police organization, the Cheka, about which we shall have more to say later. The Cheka, in scope, power, cruelty and duplicity, soon surpassed anything the czars had ever dreamed of.

One of the great intelligence services of the nineteenth century in Europe was maintained not by a government but by a private firm, the banking house of Rothschild. There was a precedent for this in the activities of a much earlier banking family, the Fuggers of Augsburg in the sixteenth century, who built up a sizable financial empire, lending money to impoverished sovereigns and states, as did the Rothschilds later. That the Fuggers made few errors in the placement of their investments was in large measure a result of the excellent private intelligence they gathered. The Rothschilds, however, once they had attained a position of some power, benefited their clients as well as themselves by their superior intelligence-gathering abilities.

In promoting their employers' financial interests from headquarters in Frankfurt-am-Main, London, Paris, Vienna and Naples, Rothschild agents were often able to gain vital intelligence before governments did. In 1815, while Europe awaited news of the Battle of Waterloo, Nathan Rothschild in London already knew that the British had been victorious. In order to make a financial killing, he then depressed the market by selling British Government securities; those who watched his every move in the market did likewise, concluding that Waterloo had been lost by the British and their allies. At the proper moment he bought back in at the low, and when the news was finally generally known, the value of government securities naturally soared.

Sixty years later Lionel Rothschild, a descendant of Nathan, on one historic evening had Disraeli as his dinner guest. During the meal a secret message came to Lionel that a controlling interest in the Suez Canal

Company, owned by the Khedive of Egypt, was for sale. The Prime Minister was intrigued with the idea, but the equivalent of about $44,000,000 was required to make the purchase. Parliament was in recess and he could not get it quickly. So Lionel bought the shares for the British Government, enabling Disraeli to pull off one of the great coups of his career. It was rumored that some of the Rothschild "scoops" were obtained by the use of carrier pigeons. There was probably little basis for the rumor, although it is true that one of the Rothschilds, immobilized in Paris when the city was surrounded by Germans in the Franco-German War of 1870, used balloons and possibly also carrier pigeons to communicate with the outside world. The world heard of the armistice ending the war through this means, rather than through conventional news channels.

The Great Powers of Europe entered World War I with intelligence services which were in no way commensurate with the might of their armed forces or equipped to cope with the complexity of the conflict to come. This was true of both sides—the Allies and the Central Powers. French military intelligence had been badly shaken up by the Dreyfus affair and was rent by internal factions and conspiracies. They calculated the size of the German Army at just half of what it was when it went into the field in 1914. The German service, which had risen to notable efficiency under Stieber in 1870, had fallen into a sad state of disrepair after his dismissal; it was moreover typical of the arrogance and self-assurance of the German General Staff of 1914 that it looked down its nose at intelligence and did not think it of importance. The Russians had achieved their great intelligence coup shortly before in the treason of the Austrian General Staff Officer, Colonel Alfred Redl, who had finally been caught in 1913. I shall have more to say of him in a later chapter. Through him they had come into possession of the Austro-Hungarian war plans, which helped them defeat the Austrians in a number of the early battles of World War I. On the other hand, the Austrians had revised some of their plans after 1913, and the Russians, blindly putting their trust in the Redl material, frequently ran into serious trouble. They also, astonishingly enough, sent military communications to their troops in the field in clear text instead of in cipher, and the Germans gleefully listened in and picked up, free of cost, valuable information about the disposition of Russian forces.

The Austrians may have balanced out Redl's treason to some extent as a result of the work of their agent, Altschiller, who was a close confidant of czarist Minister of War Vladimir A. Sukhomlinov and his wife. Sukhomlinov, a favorite of the imperial family who went out of his way to cultivate Rasputin, was notoriously vain, venal and incompetent and had the habit of leaving important military documents lying around his house. The Germans also had an agent close to this pair, a certain Colonel Myasoedev, who was supposed to be Mme. Sukhomlinov's lover, and was hanged as a spy by the Russians in 1915.

Altogether it can be said that whatever effective espionage work was accomplished during World War I, except in the tactical field, was not particularly in the area of land operations. It was chiefly in connection with naval warfare or in the remoter and peripheral areas of conflict. British competence in breaking the German naval codes was a lifesaving intelligence feat that kept Britain's head above water in the darkest days of the war. Lawrence of Arabia in the Middle East and the German, Wassmuss, in Persia performed real exploits in the fields of espionage, subversion and fomenting insurrections that truly affected the course of the war in those areas. German espionage and sabotage in the United States were among the more successful feats of their intelligence in World War I, thanks in part to our lack of preparedness with countermeasures.

World War I did, however, result in a number of innovations in espionage. One was the use of radio in wartime communications, which opened up the new possibility of gathering intelligence of immense tactical and sometimes strategic significance by intercepting radio signals and breaking codes and ciphers. The preservation of neutrality in World War I by certain strategically located countries like Sweden, Norway, Holland and Switzerland gave rise to the espionage tactic of spying on one country via a second country, despite the best efforts of the neutrals to prevent such use of their soil. This is a technique which also has been employed in peacetime, particularly in Europe. Lastly, the Far East made its first important appearance on the international espionage scene in the shape of the Japanese intelligence service, which in the ensuing years became a highly efficient and dangerous presence in the intelligence world.

The period between the two world wars saw a proliferation of intelligence services and a growing complexity in their internal structure. The targets had become increasingly technical and the world a much more complicated place. For the new dictatorships, Germany, Italy, Japan and the U.S.S.R., the intelligence service became the major instrument abroad in probing and preparing for foreign expansion. At the same time the free countries, especially England, had to take on new and enormous responsibilities in intelligence work in the face of the threat of the dictatorships. The silent warfare between the intelligence services of both sides in

World War II supplies many of the examples and case histories to which I shall refer later on. On the Allied side, in opposition to the common enemy, there was collaboration between intelligence services that is without parallel in history and which had a most welcome outcome.

During the war days when I was with OSS, I had the privilege of working with the British service and developed close personal and service relationships which remained intact after the war.

In Switzerland I made contact with a group of French officers who had maintained the tradition of the French Deuxième Bureau and who helped to build up the intelligence service of General de Gaulle and the Free French. Toward the end of the way, cooperation was established with a branch of the Italian secret service that adhered to King Victor Emmanuel when non-Fascist Italy joined the Allied cause. I also was working with the underground anti-Nazi group in the German *Abwehr*, the professional military intelligence service of the German Army. A group within the *Abwehr* secretly plotted against Hitler. The head of the *Abwehr*, the very extraordinary Admiral Canaris, was liquidated by Hitler when, following the failure of the attempt on Hitler's life in 1944, records establishing Canaris' cooperation with the plotters were discovered.

This wartime cooperation contributed, I believe, toward creating among the intelligence services of the Free World a measure of unity of purpose, and after the war a free Western Germany has made a substantial intelligence contribution. All this has helped us to counter the massive attacks which the intelligence and security services of the Communist bloc countries are making against us today.

2

The Evolution of American Intelligence

In United States history, until after World War II, there was little official government intelligence activity except in time of combat. With the restoration of peace, intelligence organizations which the stress of battle had called forth were each time sharply reduced, and the fund of knowledge and the lessons learned from bitter experience were lost and forgotten. In each of our crises, up to Pearl Harbor, workers in intelligence have had to start in all over again.

Intelligence, especially in our earlier history, was conducted on a fairly informal basis, with only the loosest kind of organization, and there is for the historian as well as the student of intelligence a dearth of coherent official records. Operations were often run out of a general's hat or a diplomat's pocket, so to speak. This guaranteed at the time a certain security sometimes lacking in later days when reports are filed in septuplicate or mimeographed and distributed to numerous officials often not directly concerned with the intelligence process. But it makes things rather difficult for the historian. At General Washington's headquarters Alexander Hamilton was one of the few entrusted with "developing" and reading the messages received in secret inks and codes, and no copies were made. Washington, who keenly appreciated the need for secrecy, kept his operations so secret that we may never have the full history of them.

To be sure, two of his intelligence officers, Boudinot and Tallmadge, later wrote their memoirs, but they were exceedingly discreet. Even forty years after the war was over, when John Jay told James Fenimore Cooper the true story of a Revolutionary spy, which the latter then used in his novel *The Spy*, Jay refused to divulge the real name of the man. Much of what we know today about intelligence in both the Revolutionary and Civil Wars was only turned up many generations after these wars were over.

Intelligence costs money, and agents have to be paid. Since it is the government's money which is being disbursed, even the most informal and swashbuckling general will usually put in some kind of chit for expenses incurred in the collection of information. Washington kept scrupulous records of money spent for the purchase of information. He generally advanced the money out of his own personal funds and then included the payment in the bill for all his expenses which he sent the Continental Congress. Since he itemized his expenses, we can see from his financial accountings that he spent around $17,000 on secret intelligence during the years of the Revolutionary War, a lot of money in those days. Walsingham, in England, two hundred years earlier, also kept such records, and it is from them that we have gleaned many of the details about his intelligence activities.

But the official accountings are not the only indicators that the pecuniary side of intelligence contributes to history. A singular attribute of intelligence work under war conditions is the delay between the completion of an agent's work and his being paid for it. He may be installed behind the enemy lines and may not get home until the war is over. Or the military unit that employed him may have moved hastily from the scene in victory or retreat, leaving him high and dry and without his reward. Thus it may happen that not until years later, and sometimes only when the former agent or his heirs have fallen on hard times, is a claim made against the government to collect payment for past services rendered. Secret intelligence being what it is, there may be no living witnesses and absolutely no record to support the claim. In any case, such instances have often brought to light intelligence operations of some moment in our own history that otherwise might have remained entirely unknown.

In December, 1852, a certain Daniel Bryan went before a justice of the peace in Tioga County, New York, and made a deposition concerning his father, Alexander Bryan, who had died in 1825. Daniel Bryan stated that General Gates in the year 1777, just before the Battle of Saratoga, had told his father that he wished him "to go into Burgoyne's Army as a spy as he wanted at that critical moment correct information as to the heft of the artillery of the enemy, the strength and number of his artillery and if possible information as to the contemplated movements of the enemy." Bryan then "went into Burgoyne's Army where he purchased a piece of cloth for a trowsers when he went stumbling about to find a tailor and thus he soon learned the strength of the artillery and the number of the Army as near as he could estimate the same and notwithstanding that the future movements of the Enemy were kept a secret, he learned that the next day the Enemy intended to take possession of Bemis heights."

The deposition goes on to tell how Alexander Bryan got away from Burgoyne's Army and reached the American lines and General Gates in time to deliver his information, with the result that Gates was on Bemis Heights the next morning "ready to receive Burgoyne's Army." As we know, the latter was soundly trounced, an action which was followed ten days later by the surrender of Burgoyne at Saratoga. According to the deposition, Bryan was never rewarded. His sick child died during the night he was away and his wife almost died too. Gates had promised to send a physician to Bryan's family, but he had never got around to it. Seventy-five years later his son put the story on record, for reasons which are still not clear, as there is no record that any claim of recompense was filed.[1]

Until accident or further research turns up additional information, we shall not know to what extent Gates' victorious strategy, which helped greatly to turn the tide of the war and was so instrumental in persuading the French to assist us, was based on the information which Bryan delivered. Sporadic finds of this kind can only make us wonder who all the other unsung heroes may have been who risked their lives to collect information for the American cause.

The one spy hero of the Revolution about whom every American schoolboy does know is, of course, Nathan Hale. Even Hale, however, despite his sacrifice, suffered comparative oblivion for decades after his death and did not become a popular figure in American history until the mid-nineteenth century. In 1799, twenty-two years after his death, an early American historian, Hannah Adams, wrote, "It is scarcely known such a character existed." In his own time, Hale's misfortune had quite a special significance for the conduct of Colonial operations.

[1]The original of this deposition is in the Walter Pforzheimer Collection on Intelligence Service through whose courtesy the above passages have been cited.

Since Hale had been a volunteer, an amateur, mightily spurred on by patriotism but sadly equipped to carry out the dangerous work of spying, his death and the circumstances of it apparently brought home sharply to General Washington the need for more professional, more carefully prepared intelligence missions. After Hale's loss, Washington decided to organize a secret intelligence bureau and chose as one of its chiefs Major Benjamin Tallmadge, who had been a classmate and friend of Nathan Hale at Yale and therefore had an additional motive in promoting the success of his new enterprise. His close collaborator was a certain Robert Townsend.

Townsend directed one of the most fruitful and complex espionage chains that existed on the Colonial side during the Revolution. At least we know of no other quite like it. Its target was the New York area, which was, of course, British headquarters. Its complexity lay not so much in its collection effort as in its communications. (I recall that General Donovan always impressed on me the vital significance of communications. It is useless to collect information unless you can quickly and accurately get it to the user.)

Since the British held New York, the Hudson and the harbor area firmly under their control, it was impossible or at least highly risky to slip through their defenses to Washington in Westchester. Information from Townsend's agents in New York was therefore passed to Washington by a highly roundabout way, which for the times, however, was swift, efficient and secure. It was carried from New York to the North Shore of Long Island, thence across Long Island Sound by boat to the Connecticut shore, where Tallmadge picked it up and relayed it to General Washington.

The best-known spy story of the Revolution other than that of Hale is the story of Major John André and Benedict Arnold. These two gentlemen might never have been discovered, in which case the damage to the patriot cause would have been incalculable, had it not been for Townsend and Tallmadge, who were apparently as sharp in the business of counterintelligence as they were in the collection of military information.

One account claims that during a visit André paid to a British major quartered in Townsend's house he aroused the suspicions of Townsend's sister, who overheard his conversation and reported it to her brother. Later, when André was caught making his way through the American line on a pass Arnold had issued him, a series of blunders which Tallmadge was powerless to prevent were instrumental in giving Arnold warning that he had been discovered, thus triggering his hasty and successful escape.

A typical "brief" written by Washington himself for Townsend late in 1778 mentioned among other things the following: ". . . mix as much as possible among the officers and refugees, visit the Coffee Houses, and all public places [in New York.]" Washington then went on to enumerate particular targets and the information he wanted about them: "whether any works are thrown up on Harlem River, near Harlem Town, and whether Horn's Hook is fortified. If so, how many men are kept at each place and what number and what sized Cannon are in those works."

This is a model for an intelligence brief. It spells out exactly what is wanted and even tells the agent how to go about getting the information.

The actual collection of information against British headquarters in New York and Philadelphia seems to have been carried out by countless private citizens, tradesmen, booksellers, tavernkeepers and the like, who had daily contact with British officers, befriended them, listened to their conversations, masquerading as Tories in order to gain their confidence, The fact that the opposing sides were made up of people who spoke the same language, had the same heritage and differed only in political opinion made spying a different and in a sense a somewhat easier task than it is in conflicts between parties of alien nationality, language and even physical aspect. By the same token, the job of counterespionage is immensely difficult under such circumstances.

One typical unsung patriot of the time was a certain Hercules Mulligan, a New York tailor with a large British clientele. His neighbors thought him a Tory or at least a sympathizer and snubbed him and made life difficult for him. On General Washington's first morning in New York after the war was over, he stopped off rather conspicuously at Mulligan's house and, to the enormous surprise of Mulligan's neighbors, breakfasted with him. After that, the neighbors understood about Mulligan. He had obviously gleaned vital information from his talkative British military customers and managed to pass it on to the General, possibly via Townsend's network.

Intelligence during the Revolution was by no means limited to military espionage in the Colonies. A fancier game of international political spying was being played for high stakes in diplomatic circles, chiefly in France, where Benjamin Franklin headed an American mission whose purpose was to secure French assistance for the Colonial cause. It was of the utmost importance for the British to learn how Franklin's negotiations were proceeding and what help the French were offering the

Colonies. How many spies surrounded Franklin and how many he himself had in England we shall probably never know. He was a careful man and he was sitting in a foreign country and he himself published little about this period of his life. However, we do know a great deal about one man who apparently succeeded in double-crossing Franklin. Or did he? That is the question.

Dr. Edward Bancroft had been born in the Colonies in Westfield, Massachusetts, but had been educated in England. He was appointed as secretary to the American commission in Paris, wormed his way into Franklin's confidence and become his "faithful" assistant and protégé for very little pay. He successfully simulated the part of a loyal and devoted American. He was able to manage nicely on his low salary from the Americans because he was being generously subsidized by the British—"£500 down, the same amount as yearly salary and a life pension." Being privy, or so he thought, to all Franklin's secret negotiations, he was no doubt a valuable agent to the British.

He passed his messages to the British Embassy in Paris by depositing them in a bottle hidden in the hollow root of a tree in the Tuileries Gardens. They were written in secret inks between the lines of love letters. Whenever he had more information than could be fitted into the bottle, or when he needed new directives from the British, he simply paid a visit to London—with Franklin's blessing, for he persuaded Franklin that he could pick up valuable information for the Americans in London. The British obligingly supplied him with what we today call "chicken feed," misleading information prepared for the opponents' consumption. Bancroft was thus one of the first double agents in our history.

To deflect possible suspicion of their agent, the British once even arrested Bancroft as he was leaving England, an action intended to impress Franklin with his bona fides and with the dangers to which his devotion to the American cause exposed him. Everything depended, of course, on the acting ability of Dr. Bancroft, which was evidently so effective that when Franklin was later presented with evidence of his duplicity he refused to believe it.

Perhaps the wily Franklin really knew of it but did not want to let on that he did. In 1777, Franklin wrote to an American lady living in France, Juliana Ritchie, who had warned him that he was surrounded with spies:

> I am much oblig'd to you for your kind Attention to my Welfare in
> the Information you give me. I have no doubt of its being well

founded. But as it is impossible to . . . prevent being watch'd by
Spies, when interested People may think proper to place them for
that purpose; I have long observ'd one rule which prevents an Incon-
venience from such Practices. It is simply this, to be concern'd in no
Affairs that I should blush to have made publick; and to do nothing
but what Spies may see and welcome. When a Man's Actions are just
and honourable, the more they are known, the more his Reputation
is increas'd and establish'd. If I was sure therefore that my Valet de
Place was a Spy, as probably he is, I think I should not discharge him
for that, if in other Respects I lik'd him.

—B.F.[2]

Once when the British lodged an official diplomatic protest with the
French regarding the latter's support of the American cause, they based
the protest on a secret report of Bancroft's, quoting facts and figures he
had received from Franklin and even using Bancroft's wording, a bit of
a slip that happens from time to time in the intelligence world. Bancroft
was mortally afraid that Franklin might smell a rat and suspect him. He
even had the British give him a passport so that he could flee on a moment's notice if necessary. Franklin did express the opinion on this occasion that "such precise information must have come from a source
very near him," but as far as we know he did nothing else about it.

The British, also, had reason to suspect Bancroft. George III does
not seem to have fully trusted him or his reports since he caught him out
investing his ill-gotten pounds in securities whose value would be en-
hanced by an American victory.

Bancroft's duplicity was not clearly established until 1889, when certain papers in British archives pertaining to the Revolutionary period were
made public. Among them, in a letter addressed to Lord Carmarthen, Secretary of State for Foreign Affairs, and written in 1784, Bancroft set down
in summary form his activities as a British agent. It seems the British government had fallen behind in their payments to him and Bancroft was putting in a claim and reminding his employers of his past services. He closed
with the words: "I make no Claim beyond the permanent pension of £500
pr an. for which the Faith of Government has often been pledged; and for
which I have sacrificed near eight years of my life."

[2]The original of this letter is in the collection of Franklin papers of the American Philosophical Society in Philadelphia.

Franklin's own agents in London were apparently highly placed. Early in 1778 Franklin knew the contents of a report General Cornwallis submitted in London on the American situation less than a month after Cornwallis had delivered it. The gist of the report was that the conquest of America was impossible. If Franklin's agents had penetrated the British government at this level, it is possible that they had caught wind of the intelligence Bancroft was feeding the British.

In the Civil War, even more than in the Revolution, the common heritage and language of the two parties to the conflict and the fact that many people geographically located on one side sympathized with the political aims of the other made the basic task of espionage relatively simple, while making the task of counterespionage all the more difficult. Yet the record seems to show that few highly competent continuous espionage operations, ones that can be compared in significance of achievement and technical excellence with those of the Revolution, existed on either side. No great battles were won or lost or evaded because of superior intelligence. Intelligence operations were limited for the most part to more or less localized and temporary targets. As one writer has put it, "There was probably more espionage in one year in any medieval Italian city than in the four-year War of Secession."

The reasons for this are numerous. There was no existing intelligence organization on either side at the outbreak of the war nor was there any extensive intelligence experience among our military personnel of that day. Before the Revolution, the Colonial leaders had been conspiring and carrying out a limited secret war against the British for years and by the time of open conflict had a string of active "sources" working for them in England and moreover possessed tested techniques for functioning in secret at home. This was not the case in the North or the South before the Civil War. Washington was an outstandingly gifted intelligence chief. He himself directed the entire intelligence effort of the American forces, even to taking a hand personally in its more important operations. There was no general with a similar gift in the whole galaxy of Federal or Confederate generals. Lastly, the Civil War by its very nature was not a war of surprises and secrets. Large lumbering armies remained encamped in one place for long periods of time, and when they began to move word of their movements spread in advance almost automatically. Washington, with far smaller numbers of men could plant false information as to his strength and could move his troops so quickly that a planned British action wouldn't find them where they had been the day before, especially when Washington through his networks knew in advance of the British move.

At the beginning of the Civil War the city of Washington was a sieve and the organization on the Northern side so insecure that the size and movements of its forces were apparent to any interested observer. It has been said that the Confederate side never again had such good intelligence to help them as they did at the opening Battle of Bull Run.

One of the first events of the period which apparently pointed up the need for a secret intelligence service was the conspiracy of a group of hotheads in Baltimore to assassinate Lincoln on the way to his first inauguration in February, 1861. Allan Pinkerton, who had already achieved some fame working as a private detective for the railroads, had been hired by some of Lincoln's supporters to protect him. Pinkerton got Lincoln to Washington without incident by arranging to have the presidential train pass through Baltimore unannounced late at night. At the same time Pinkerton's operatives "penetrated" the Baltimore conspirators and kept a close watch on their activities.

Good as Pinkerton was at the job of security and counterespionage, he had little to recommend him for the work of intelligence collection except for one excellent agent, a certain Timothy Webster, who produced some good information entirely on his own in Virginia. Unfortunately, Webster was captured early in the war, thanks to a foolish maneuver of Pinkerton, and was subsequently executed. We next find Pinkerton working directly with General McClellan on military intelligence and right in the General's headquarters. Pinkerton's idea of military intelligence was to count the noses of the opposing troops and then to count them all over again to be sure the first figure was right. Since McClellan was famous for not going into battle unless he commanded overwhelming numbers, it is not likely that Pinkerton's nose-counting contributed significantly to the outcome of any battle. Even with overwhelming odds in his favor, McClellan was outmaneuvered by Lee at Antietam. When Lincoln removed him from his command after this battle, Pinkerton resigned, leaving the Union virtually without a secret service.

The fact that Lincoln had hired an agent of his own on a military intelligence mission at the time of the Battle of Bull Run did not come to light until 1876, and then, as so often is the case, it was revealed in the form of a claim against the government for reimbursement. In March of 1876, the United States Supreme Court heard a case on appeal from the U.S. Court of Claims in which a certain Enoch Totten brought a claim against the government "to recover compensation for services alleged to have been rendered" by a certain William A. Lloyd, "under contract with President Lincoln, made in July 1861, by which he was to proceed South and ascertain the number of troops stationed at different points in the insurrectionary States, procure plans of forts and fortifications . . . and report the facts to the President. . . . Lloyd proceeded . . . within the rebel lines, and remained there during the entire period of the war, collecting and from time to time transmitting information to the President." At the end of the war he had been paid his expenses but not the salary of $200 a month which Lincoln, according to the claim, had promised him. The case itself is interesting even with only these meager facts because of the light it casts on Lincoln's foresight at this time and the security with which he must have handled the matter throughout the four long years of the war. As the Supreme Court stated in its opinion: "Both employer and agent must have understood that the lips of the other were to be forever sealed respecting the relation of either to the matter."

Also, this case established the precedent that an intelligence agent cannot recover by court action against the government for *secret* service rendered. Said the Court: "Agents . . . must look for their compensation to the contingent fund of the department employing them, and to such allowance from it as those who dispense the fund may award. The secrecy which such contracts impose precludes any action for this enforcement." This is a warning to the agent that he had better get his money on the barrelhead at the time of his operation.

After Pinkerton left the scene, an effort was made to create a purely military intelligence organization known as the Bureau of Military Information. The responsibility for it was assigned to Major (later General) George H. Sharpe, who appears to have been a fair-to-middling bureaucrat but is not known to have conceived or mounted significant intelligence operations on his own. However, good information was brought to the Union forces by occasional brave volunteers, most of whom generated their own operations and communications without good advice from anybody. One of these was Lafayette Baker, who posed as an itinerant photographer in the South and made a specialty of visiting Confederate camps in Virginia, taking pictures of the soldiers stations in them, at the same time gathering valuable military information. He later rose to brigadier general and took charge of the National Detective Police, a sort of precursor of today's secret service. Where Pinkerton had excelled at counterespionage but had little to recommend him as an espionage operator, Baker excelled in the latter craft, but his failures as a chief of secret service lost us one of our greatest Presidents. To this day, no one knows where Baker's men were on the night of April 14, 1865, when Abraham Lincoln was sitting in an unguarded box watching a play in Ford's Theater, or why the assassins who gathered at Mrs. Suratt's boardinghouse, whose fanatical opinions were well known throughout Washington, were not being watched by Baker. Nor was the capture of Booth and his accomplices the work of Baker, although he took credit for it.

Elizabeth van Lew, another volunteer in the South and a resident of Richmond, stayed at her post throughout the entire war and is accounted the single most valuable spy the North ever had. Grant himself stated that she had sent the most valuable information received from Richmond during the war. In Civil War espionage any "penetration" of an important headquarters, always the most dramatic high-level intelligence operations, is conspicuously missing, as are most of the more rewarding and devious undertakings of espionage. The closest thing to it, however, is alleged to have been achieved by Elizabeth van Lew when she procured a job for one of her Negro servants as a waitress in the house of Jefferson Davis, transmitting the intelligence this produced to Major Sharpe in Washington.

In the 1880s the first permanent peacetime military and naval intelligence organizations were created in the United States. The Army unit was known as the Military Information Division and came under the Adjutant General's Office. The Navy's Office of Intelligence, founded in 1882, first belonged to the Bureau of Navigation. During the same decade the first U.S. military and naval attachés were posted to our embassies and legations abroad, where they were to function as observers and intelligence officers.

Elbert Hubbard's once-popular tale *A Message to Garcia* immortalized an exploit of American intelligence during the Spanish-American War that might otherwise have been forgotten. Actually, Hubbard got the story backward. The usual point of an intelligence mission is to get the needed information *to* headquarters *from* a target area. The Lieutenant Rowan of Hubbard's story was, in real life, supposed to reach Garcia, which was not easy, but his chief purpose was to get information from Garcia about the disposition of Spanish troops and then bring it back. Obviously the latter part of the mission was more important that the former.

It is worth recalling that the man who dispatched Rowan on his mission, Col. Arthur L. Wagner, was one of the pioneers of American intelligence and even wrote a book on the subject. When he was assigned in 1898 to the Cuban Expeditionary Force as commander of the "Department of Intelligence in the Field," General Shafter, at the head of this force, would have none of any such newfangled notions and refused to accept him. At the time of Wagner's death in 1905, his commission as a brigadier general was lying on the President's desk for signature. Wagner, like many of our earlier intelligence officers, was born a little too soon.

Since the 1880s also saw the founding of our Naval Intelligence, the Spanish-American War was the occasion for certain important intelligence exploits of our Navy. An unusual and romantic account has been preserved in the Navy's archives which tells the story of two young American ensigns who, disguising themselves as Englishmen and traveling under assumed names, went to Spain and Spanish-held territories to watch and report on movements of the Spanish fleet. They kept an eye on Admiral Cervera's ships and followed them from Cadiz and Gibraltar all the way to Puerto Rico, and "several times narrowly escaped detection."

In 1903, with the creation of an Army General Staff, the Military Information Division was incorporated into it as the "Second Division," thus beginning the tradition of G-2, which has since remained the designation for intelligence in the American Army. This early G-2, however, from lack of interest and responsibility dwindled almost to the point of disappearance, with the result that World War I found us again without any real intelligence service. But this time our situation was different. We were fighting abroad, the whole period during which our troops were directly engaged lasted little over a year, and we had allies. There was no time to develop a full-fledged intelligence arm nor did we have to, since we could rely largely on the British and French for military intelligence and particularly for order of battle.

But we learned rapidly—due largely to a group of officers to whom I wish to pay tribute. There was, first of all, Colonel Ralph H. Van Deman, who is considered by many to be the moving force in establishing a U.S. military intelligence. His work is described in what I consider the best account by an American author of intelligence services through the ages, *The Story of Secret Service*, by Richard Wilmer Rowan. I worked personally with Colonel Van Deman in World War I when I was in Bern, and I can attest to the effective work that he and his successors and their naval opposite numbers did in building up the basis of our military intelligence today. But in peacetime they had far too little support in the military services.

By the time the war was over, the basic framework had been established for the various military and naval intelligence branches which continued to exist, even though in skeleton form, until the outbreak of the Second World War—G-2, CIC (Counter Intelligence Corps, which until 1942 was called the Corps of Intelligence Police) and ONI (Office of Naval Intelligence). Of equal importance was our initial experience during World War I in the field of cryptography, of which I shall have more to say in a later chapter. In this area, too, a skeleton force working during the interim years of peace succeeded in developing the most vital instrument of intelligence which we possessed when we were finally swept into war again in 1941—the ability to break the Japanese diplomatic and naval codes.

It was only in World War II, and particularly after the Pearl Harbor attack, that we began to develop, side by side with our military intelligence organizations, an agency for secret intelligence collection and operations. As I mentioned earlier, the origin of this agency was a summons by President Franklin D. Roosevelt to William J. Donovan in 1941 to come down to Washington and work on this problem.

Donovan was eminently qualified for the job. A distinguished lawyer, a veteran of World War I who had won the Medal of Honor, he had divided his busy life in peacetime between the law, government service and politics. He knew the world, having traveled widely. He understood people. He had a flair for the unusual and for the dangerous, tempered with judgment. In short, he had the qualities to be desired in an intelligence officer.

The Japanese sneak attack on Pearl Harbor and our entry into the war naturally stimulated the rapid growth of the OSS and its intelligence operations.

It had begun, overtly, as a research and analysis organization, manned by a hand-picked group of some of the best historians and other scholars available in this country. By June, 1942, the COI (Coordinator of Information), as Donovan's organization had been called at first, was renamed the Office of Strategic Services (OSS) and told "to collect and analyze strategic information and to plan and operate special services."

By this time the OSS was already deep in the task of "special services," a cover designation for secret intelligence and secret operations of every conceivable character among which the support of various anti-Nazi underground groups behind the enemy lines and covert preparations for the invasion of North Africa were perhaps most significant.

During 1943, elements of the OSS were at work on a world-wide basis, except for Latin America, where the FBI was operating, and parts of the Far Eastern Command, which General MacArthur had already pre-empted.

Its guerilla and resistance branch, modeled on the now well-publicized British Special Operations Executive (SOE) and working closely with the latter in the European Theater, had already begun to drop teams of men and women into France, Italy and Yugoslavia and in the China-Burma-India Theater of war. The key idea behind these operations was to support, train and supply already existing resistance movements or, where there were none, to organize willing partisans into effective guerilla units. The Jedburghs, as they were called, who dropped into France, and Detachment 101, the unit in Burma, were among the most famous of these groups. Later the OSS developed special units for the creation and dissemination of black propaganda, for counterespionage, and for certain sabotage and resistance tasks that required unusual talents, such as underwater demolitions or technical functions in support of regular intelligence tasks. In conjunction with all these undertakings, it had to develop its own training schools.

Toward the end of the war, as our armies swept over Germany, it created special units for the apprehension of war criminals and the recovery of looted art treasures as well as for tracking down the movements of funds which, it was thought, the Nazi leaders would take into hiding in order to make a comeback at a later date. There was little that it did not attempt to do at some time or place between 1942 and the war's end.

For a short time after V-J Day, it looked as though the U.S. would gradually withdraw its troops from Europe and the Far East. This would probably have included the disbanding of intelligence operations. In fact, it seemed likely at the end of 1945 that we would do what we did after World War I—fold our tents and go back to business-as-usual. But this time, in contrast to 1919 when we repudiated the League of Nations, we became a charter member of the United Nations and gave it our support in hopes that it would grow up to be the keeper of world peace.

If the Communists had not overreached themselves, our government might well have been disposed to leave the responsibility for keeping the peace more and more to the United Nations. In fact, at Yalta Stalin asked President Roosevelt how long we expected to keep our troops in Europe.

The President answered, not more than two years. In view of the events that took place in rapid succession during the postwar years, it is clear that in the period between 1945 and 1950 Premier Stalin and Mao Tse-tung decided that they would not wait for us to retire gracefully from Europe and Asia; they would kick us out.

Moscow installed Communist regimes in Poland, Rumania and Bulgaria before the ink was dry on the agreements signed at Yalta and Potsdam. The Kremlin threatened Iran in 1946, and followed this in rapid succession by imposing a Communist regime on Hungary, activating the civil war in Greece, staging the takeover of Czechoslovakia and instituting the Berlin blockade. Later, in 1950, Mao joined Stalin to mastermind the attack on South Korea. Meanwhile, Mao had been consolidating his position on the mainland of China. These blows in different parts of the world aroused our leaders to the need for a worldwide intelligence system. We were, without fully realizing it, witnessing the first stages of a master plan to shatter the societies of Europe and Asia and isolate the United States, and eventually to take over the entire world. What we were coming to realize, however, was the need to learn a great deal more than we knew about the secret plans of the Kremlin to advance the frontiers of Communism.

In his address to Congress on March 12, 1947, President Truman declared that the security of the country was threatened by Communist actions and stated that it would be our policy "to help free peoples to maintain their free institutions and their national integrity against aggressive movements seeking to impose on them totalitarian regimes." He added that we could not allow changes in the status quo brought about by "coercion or by such subterfuges as political infiltration," in violation of the United Nations Charter.

It was by then obvious that the United Nations, shackled by the Soviet veto, could not play the role of policeman. It was also clear that we had a long period of crisis ahead of us. Under these conditions, a series of measures were taken by the government to transform our words into action. One of the earliest was the reorganization of our national defense structure, which provided for the unification of the military services under a Secretary of Defense and the creation of the National Security Council.

At that time President Truman recommended that a central intelligence agency be created as a permanent agency of government. A Republican Congress agreed and, with complete bipartisan approval, the CIA was established in the National Security Act of 1947. It was an openly acknowledged arm of the executive branch of government, although, of course, it had many duties of a secret nature. President Truman saw to it that the new agency was equipped to support our government's effort to meet Communist tactics of "coercion, subterfuge, and political infiltration." Much of the knowhow and some of the personnel of the OSS were taken over by the Central Intelligence Agency.

The two years between the end of World War II when the OSS was dissolved and the creation of CIA in the fall of 1947 had been a period of interdepartmental infighting as to what to do with intelligence. Fortunately, many experienced officers of the OSS remained on during this period in the various intelligence units which functioned under the aegis of the State and War Departments in the postwar period.

This was largely due to the foresight of General Donovan. At an early date he had directed President Roosevelt's attention to the importance of preserving the OSS assets and providing for the carrying on of certain of the intelligence functions which had devolved upon the OSS during World War II.

As early as October, 1944, Donovan had discussed this whole problem with the President, and in response to his request had sent him a memorandum outlining his ideas of what an intelligence service should be equipped to do in the postwar period. In this memorandum he stressed that while intelligence operations during the war were mainly in support of the military and hence had been placed under the Joint Chiefs of Staff, in the postwar period he felt they should be placed under the direct supervision of the President. He further proposed that a central intelligence authority, to include the Secretaries of State and Defense, as well as a representative of the President himself, should be created to supervise and coordinate intelligence work. In concluding his memorandum, General Donovan stated: "We have now in government the trained and specialized personnel needed for the task. This talent should not be dispersed."

Under the pressure of events during the last months of the war, it was not until April 5, 1945, that President Roosevelt, as one of his last acts, answered General Donovan's memorandum. The President instructed him to call together "the chiefs of foreign intelligence and internal security units in the various Executive agencies so that a consensus of opinion can be secured" as "to the proposed centralized Intelligence service."

President Truman took the oath of office on April 12, 1945, and was of course immediately involved in all of the intricate questions arising out of the end of the war in Europe, the prosecution of the war against

Japan and the preparation for the Potsdam Conference of July, 1945. But on April 26 he had a chance to discuss intelligence with the Director of the Bureau of the Budget, Harold D. Smith. He had got into the act in connection with the preparation of the new budget and had his own ideas about how intelligence should be organized. He had already sent President Roosevelt a memorandum, in which he pointed out, as President Truman reports,[3] "that a tug of war was going on among the FBI, the Office of Strategic Services, the Army and Navy Intelligence, and the State Department." President Truman added in his memoirs:

> I considered it very important to this country to have a sound, well-organized intelligence system, both in the present and in the future. Properly developed, such a service would require new concepts as well as better-trained and more competent personnel. Smith suggested, and I agreed, that studies should be undertaken at once by his specially trained experts in this field. Plans needed to be made, but it was imperative that we refrain from rushing into something that would produce harmful and unnecessary rivalries among the various intelligence agencies. I told Smith that one thing was certain—this country wanted no Gestapo under any guise or for any reason.

For the next few months the issue was hotly debated, with the Joint Chiefs of Staff playing an important role. They instructed their Joint Intelligence Committee, on which all the military and civilian foreign Intelligence agencies, including OSS, were represented, to study the proposals Donovan had earlier submitted to President Roosevelt, as well as those of other interested agencies.

Meanwhile the Bureau of the Budget continued its own activities and prepared an Executive Order for President Truman's signature putting the Office of Strategic Services into liquidation. When the Joint Chiefs heard of this, they urged the President to defer action until their views could be presented. However, this word reached the White House too late. The President, on the 20th of September, 1945, issued an Executive Order providing for the termination of the OSS and placing its research unit in the Department of State and the other remaining units under the Secretary of War. These latter were put together in

[3]Harry S. Truman, *Years of Decision* (New York: Doubleday & Co., Inc., 1955), p. 98.

an organization known as the Strategic Services Unit (SSU). SSU was not combined with G-2 but was put under the Under Secretary of War, and it is only fair to say that throughout the ensuing struggle for control and until SSU was taken over by the Central Intelligence Group (CIG), SSU was left largely autonomous in its operations and received complete administrative support from the Army.

The tug of war had continued between the Department of State, which desired to take over the postwar leadership of foreign intelligence, and the military services, including the Joint Chiefs of Staff, which wished to continue the domination they had exercised during the war.

To help resolve these conflicts of interest, the President called on an old friend, Sidney W. Souers, who had been serving the Navy Department in an intelligence capacity. He had been promoted to flag rank in 1945 and made Deputy Chief of Naval Intelligence. Souers worked closely with Admiral Leahy and James S. Lay, Jr., who had been secretary of the JIC and later became Executive Secretary of the National Security Council.

Of the many studies and proposals, probably the most influential was that of the so-called Lovett Committee, headed by Robert A. Lovett, Assistant Secretary of War for Air. This contemplated a Central Intelligence Agency supported by an independent budget which would be responsible only to a National Intelligence Authority composed of the Secretaries of State, War and Navy and a representative of the Joint Chiefs of Staff.

Finally, on January 22, 1946, President Truman reached his own decision and acted. In a directive to the Secretaries of State, War and Navy, he ordered that they, together with a personal representative of the President (Admiral William Leahy became the President's designee), should constitute themselves as the National Intelligence Authority. This was to supervise the new intelligence organization which was placed under a director of central intelligence. Admiral Souers was appointed the first head of the new agency, known as the Central Intelligence Group (CIG). He resigned six months later, but continued as an adviser to his successor, General Hoyt Vandenberg.

Later, President Truman, using his directive of January 22 and the experience gained through the operations of the CIG, approved the legislation creating the Central Intelligence Agency and included it in the National Security Act of 1947.

Under the Act, the Central Intelligence Agency was placed under the direction of the National Security Council, which is composed of the President, the Secretary of State, the Secretary of Defense and other primary

Presidential advisers in the field of foreign affairs. Interestingly enough, CIA is the sole agency of government which as a matter of law is under the National Security Council, whose function is solely to advise the President. Thus there was firmly established the principle of control of intelligence at the White House level, which President Truman had developed in creating the National Intelligence Authority.

The CIA was not patterned wholly either on the OSS or on the structural plan of earlier or contemporary intelligence organizations of other countries. Its broad scheme was in a sense unique in that it combined under one leadership the overt task of intelligence analysis and coordination with the work of secret intelligence operations of the various types I shall describe. Also, the new organization was intended to fill the gaps in our existing intelligence structure without displacing or interfering with other existing U.S. intelligence units in the Departments of State and Defense. At the same time, it was recognized that the State Department, heretofore largely dependent for its information on the reports from diplomatic establishments abroad, and the components of the Defense Department, relying mainly on attachés and other military personnel abroad, could not be expected to collect intelligence on all those parts of the world that were becoming increasingly difficult of access nor to groom a standing force of trained intelligence officers. For this reason, CIA was given the mandate to develop its own secret collection arm, which was to be quite distinct from that part of the organization that had been set up to assemble and evaluate intelligence from all parts of the government.

One of the unique features of CIA was that its evaluation and coordinating side was to treat the intelligence produced by its clandestine arm in the same fashion that information from other government agencies was treated. Another feature of the CIA's structure, which did not come about all at once but was the result of gradual mergers which experience showed to be practical and efficient, was the incorporation of all clandestine activities under one roof and one management. Traditionally, intelligence services have kept espionage and counterespionage in separate compartments and all activities belonging in the category of political of psychological warfare in still another compartment. CIA abandoned this kind of compartmentalization, which so often leads to neither the right hand nor the left knowing what the other is doing.

The most recent development in American intelligence has been a unification of the management of the various intelligence branches of the armed forces. In August, 1961, the Defense Intelligence Agency

(DIA) was established under a directive issued by the Department of Defense. An outstanding Air Force officer, Lieutenant General Joseph F. Carroll, was named as its first director. His first deputy director, Lieutenant General William W. Quinn, and I worked closely together when Quinn was the very able G-2 to General Alexander M. Patch of the Seventh Army during the invasion of Southern France and Germany. In those days, in the summer and autumn of 1944, I used to meet secretly with Quinn at points in liberated France near the northern Swiss border and supply him with all the military intelligence I could gather on Nazi troop movements and plans as Hitler's forces retreated toward the mountain "redoubt" of Southern Germany and Austria. Rear Admiral Samuel B. Frankel, the Chief of Staff, likewise an experienced intelligence officer, made a special contribution to the work of the United States Intelligence Board (USIB) during the years when I served it as chairman. DIA was not a merger of the intelligence branches of the armed services, but primarily an attempt to achieve maximum coordination and efficiency in the intelligence processes of the three services. On February 1, 1964, the Department of Defense issued a comprehensive directive establishing intelligence career programs to create a broad professional base of trained and experienced intelligence officers.

Thus, in contrast to our custom in the past of letting the intelligence function die when the war was over, it has been allowed to grow to meet the ever-widening and more complex responsibilities of the time. The formation of such agencies as the DIA, like the earlier creation of CIA itself, is the result of studied effort to give intelligence its proper stature in our national security structure. There is, of course, always the possibility that two such powerful and well-financed agencies as CIA and DIA will become rivals and competitors. There is obviously also room here for an expansion of traditional Army ambitions to run a full-fledged and independent covert collection service of its own, which is hardly justifiable under present circumstances. It could also be both expensive and dangerous. A clear definition of functions is always a requisite. In broad outlines, this already exists. Furthermore, the high caliber of the officers, military and civilian, directing the two agencies, if maintained, should guarantee effective performance, but it is vital to protect the authority of the Director of Central Intelligence to coordinate the work of foreign intelligence, under the President, and to see to the preparation of our National Intelligence estimates, which I shall describe in detail later.

3

America's Intelligence Requirements

Intelligence is probably the least understood and the most misrepresented of the professions. One reason for this was well expressed by President Kennedy when, on November 28, 1961, he came out to inaugurate the new CIA Headquarters Building and to say good-bye to me as Director. He then remarked: "Your successes are unheralded, your failures are trumpeted." For obviously you cannot tell of operations that go along well. Those that go badly generally speak for themselves.

The President then added a word of encouragement to the several thousand men and women of CIA:

> . . . but I am sure you realize how important is your work, how essential it is—and in the long sweep of history how significant your efforts will be judged. So I do want to express my appreciation to you now, and I am confident that in the future you will continue to merit the appreciation of our country, as you have in the past.

It is hardly reasonable to expect proper understanding and support for intelligence work in this country if it is only the insiders, a few people within the executive and legislative branches, who know anything whatever about the CIA. Others continue to draw their knowledge from the so-called inside stories by writers who have never been on the inside.

There are, of course, sound reasons for not divulging intelligence secrets. It is well to remember that what is told to the public also gets to the enemy. However, the discipline and techniques—what we call the tradecraft of intelligence—are widely known in the profession, whatever the nationality of the service may be. What must not be disclosed, and will not be disclosed here, is where and how and when the tradecraft has been or will be employed in particular operations unless this has already been disclosed elsewhere, as in the case of the U-2, for example.

CIA is not an underground operation. All one needs to do is to read the law—the National Security Act of 1947—to get a general idea of what it is set up to do. It has, of course, a secret side, and the law permits the National Security Council, which in effect means the President, to assign to the CIA certain duties and functions in the intelligence field in addition to those specifically enumerated in the law. These functions are not disclosed. But CIA is not the only government agency where secrecy is important. The Departments of State and of Defense also guard with great care the security of much that they do.

One of my own guiding principles in intelligence work when I was Director of Central Intelligence was to use every human means to preserve the secrecy and security of those activities, but only those where this was essential, and not to make a mystery of what is a matter of common knowledge or obvious to friend and foe alike.

Shortly after I became Director, I had a good illustration of the futility of certain kinds of secrecy. Dr. Milton Eisenhower, brother of the President, had an appointment to see me. The President volunteered to drop him by at my office. They started out (I gather without forewarning to the Secret Service), but could not find the office until a telephone call was put through to me for precise directions. This led me to investigate why all this futile secrecy. At that time the CIA Headquarters bore at the gate the sign "Government Printing Office." However, Washington sightseeing bus drivers made it a practice to stop outside our front gate. The guide would then harangue the occupants of the bus with information to the effect that behind the barbed wire they saw was the most secret, the most concealed place in Washington, the headquarters of the American spy organization, the Central Intelligence Agency. I also found out that practically every taxicab driver in Washington knew the location. As soon as I put up a proper sign at the door, the glamour and mystery disappeared. We were no longer either sinister or mysterious to visitors to the Capital; we became just another government office. Too much secrecy can be self-defeating just as too much talking can be dangerous.

An instance where a certain amount of publicity was helpful in the collection of intelligence occurred during World War II when I was sent to Switzerland for General Donovan and the OSS in November of 1942. I had a position in the American Legation as an assistant to the Minister. One of the leading Swiss journals produced the story that I was coming there as a secret and special envoy of President Franklin D. Roosevelt. Offhand one might have thought that this unsought advertisement would have hampered my work. Quite the contrary was the case. Despite my modest but truthful denials of the story, it was generally believed. As a result, to my network flocked a host of informants, some cranks, it is true, but also some exceedingly valuable individuals. If I could not separate the wheat from the chaff with only a reasonable degree of error, then I was not qualified for my job, because the ability to judge people is one of the prime qualities of an intelligence officer.

When we try to make a mystery out of everything relating to intelligence, we tend to dissipate our effort to maintain the security of operations where secrecy is essential to success. Each situation has to be considered according to the facts, keeping in mind the principle of withholding from a potential enemy all useful information about secret intelligence operations or personnel engaged in them. The injunction that George Washington wrote to Colonel Elias Dayton on July 26, 1777, is still applicable to intelligence operations today:

> The necessity of procuring good Intelligence is apparent and need not be further urged. All that remains for me to add, is, that you keep the whole matter as secret as possible. For upon Secrecy, Success depends in most Enterprizes of the kind, and for want of it, they are generally defeated, however well planned and promising a favourable issue.[1]

On the whole, Americans are inclined to talk too much about matters which should be classified. I feel that we hand out too many of our secrets, particularly in the field of military "hardware" and weaponry, and that we often fail to make the vital distinction between the types of operation that should be secret and those which, by their very nature, are not and cannot be kept secret. There are times when our press is overzealous in seeking "scoops" with regard to future diplomatic, political and military moves. We have learned the importance of secrecy in

[1]Pforzheimer Collection.

time of war, although even then there have been serious indiscretions at times. But it is well to recognize that in the Cold War our adversary takes every advantage of what we divulge or make publicly available.

To be sure, with our form of government, and in view of the legitimate interest of the public and the press, it is impossible to erect a wall around the whole business of intelligence, nor do I suggest that this be done. Neither Congress nor the executive branch intended this when the law of 1947 was passed. Furthermore, certain information must be given out if public confidence in the intelligence mission is to be strengthened and if the profession of the intelligence officer is to be properly appreciated.

Most important of all, it is necessary that both those on the inside—the workers in intelligence—and the public should come to share in the conviction that intelligence operations can help mightily to protect the nation.

In our time, the United States is being challenged by a hostile group of nations that profess a philosophy of life and of government inimical to our own. This in itself is not a new development; we have faced such challenges before. What has changed is that now, for the first time, we face an adversary possessing the military power to mount a devastating attack directly upon the United States, and in the era of nuclear missiles this can be accomplished in a matter of minutes or hours with a minimum of prior alert.

To be sure, we possess the same power against our adversary. But in our free society our defenses and deterrents are largely prepared in an open fashion, while our antagonists have built up a formidable wall of secrecy and security. In order to bridge this gap and help to provide for strategic warning, we have to rely more and more upon our intelligence operations.

The Departments of State and Defense are collecting information abroad, and their intelligence experts are analyzing it, preparing reports and doing a good job of it. Could they not do the whole task?

The answer given to this question fifteen years ago by both the executive and legislative branches of our government was "No." Underlying this decision was our growing appreciation of the nature of the Communist menace, its self-imposed secrecy and the security measures behind which it prepares its nuclear missile threat and its subversive penetration of the Free World.

Great areas of both the Soviet Union and Communist China are sealed off from foreign eyes. These nations tell us nothing about their military establishments that is not carefully controlled, and yet such knowledge is needed for our defense and for that of the Free World. They reject the principle of inspection which we have considered essential to a controlled disarmament. They boldly proclaim that this secrecy is a great asset and a basic element of policy. They claim the right to arm in secret so as to be able, if they desire, to attack in secret. They curtly refused the "open sky" proposal of President Eisenhower in 1955, which we were prepared to accept for our country if they would for theirs. This refusal has left to intelligence the task of evening the balance of knowledge and hence of preparation by breaking through this shield of secrecy.

The Berlin Wall not only shut off the two halves of a politically divided city from each other and limited the further escape of East Germans to the West in any appreciable number. It also tried to plug one of the last gaps in the Iron Curtain—that barrier of barbed wire, land mines, observation towers, mobile patrols and sanitized border areas stretching southward from the Baltic. When they put up the Berlin Wall, the Soviets finished sealing off Eastern Europe in their fashion, and it took them sixteen years to do it.

Yet there are ways of getting under or over, around or even through this barrier. It is just the first of a series of obstacles. Behind that first wall, there are further segregated and restricted areas and, behind these, the walls of institutional and personal secrecy which all together protect everything the Soviet state believes could reveal either strength or weakness to the inquisitive West.

The Iron and Bamboo Curtains divide the world in the eyes of Western intelligence into two kinds of places—free areas and "denied areas." The major targets lie in the denied areas behind the curtains. These are the military, technical, industrial and nuclear installations that constitute the backbone of Communist power—the capabilities. These are also the plans of the people who guide Soviet Russia and Communist China—their war-making intentions and their "peaceful" political intentions.

Against these targets the overt intelligence collection work of the State and Defense Departments, though of great value, is not enough. The special techniques which are unique to secret intelligence operations are needed to penetrate the security barriers of the Communist bloc.

Today's intelligence service also finds itself in the situation of having to maintain a constant watch in every part of the world, no matter what may at the moment be occupying the main attention of diplomats and military men. Our vital interests are subject to attack in almost every quarter of the globe at any time.

A few decades ago no one would have been able or willing to predict that in the 1960s our armed forces would be stationed in Korea and be deeply engaged in South Vietnam, that Cuba would have become a hostile Communist state closely allied with Moscow, or that the Congo would have assumed grave importance in our foreign policy. Yet these are all facts of life today. The coming years will undoubtedly provide equally strange developments.

Today it is impossible to predict where the next danger spot may develop. It is the duty of intelligence to forewarn of such dangers, so that the government can take action. No longer can the search for information be limited to a few countries. The whole world is the arena of our conflict. In this age of nuclear missiles, even the Arctic and the Antarctic have become areas of strategic importance. Distance has lost much of its old significance, while time, in strategic terms, is counted in hours or even minutes. The oceans, which in World War II still protected this country and allowed it ample time to prepare, are as broad as ever. But now they can be crossed by missiles in a matter of minutes and by bombers in a few hours. Today the United States is in the front line of attack, for it is the prime target of its adversaries. No longer does an attack require a long period of mobilization with its telltale evidence. Missiles stand ready on their launchers, and bombers are on the alert.

Therefore an intelligence service today has an additional responsibility, for it cannot wait for evidences of the likelihood of hostile acts against us until after the decision to strike has been made by another power. Our government must be both forewarned and forearmed. The situation becomes all the more complicated when, as in the case of Korea and Vietnam, a provocative attack is directed not against the U.S. but against some distant overseas area which, if lost to the Free World, would imperil our own security. A close-knit, coordinated intelligence service, continually on the alert, able to report accurately and quickly on developments in almost any part of the globe, is the best insurance we can take out against surprise.

The fact that intelligence is alert, that there is a possibility of forewarning, could itself constitute one of the most effective deterrents to a potential enemy's appetite for attack. Therefore the fact that such a weapon of warning can be created should not be kept secret but should be made well known, though the means and mechanics of warning should remain secret. Intelligence should not be a taboo subject. What we are striving to achieve and have gone far toward achieving—the most effective intelligence service in the world—should be an advertised fact.

In addition to getting the information, there is also the question of how it should be processed and analyzed. I feel that there are important reasons for placing the responsibility for the preparation and coordination of our intelligence analyses with a centralized agency of government which has no responsibility for policy or for choosing among the weapons systems which will be developed for our defense. Quite naturally policymakers tend to become wedded to the policy for which they are responsible, and State and Defense employees are no exception to this very human tendency. They are likely to view with a jaundiced eye intelligence reports that might tend to challenge existing policy decisions or require a change in cherished estimates of the strength of the Soviets in any particular military field. The most serious occupational hazard we have in the intelligence field, the one that causes more mistakes than any foreign deception or intrigue, is prejudice. I grant that we are all creatures of prejudice, including CIA officials, but by entrusting intelligence coordination to our central intelligence service, which is excluded from policymaking and is married to no particular military hardware, we can avoid, to the greatest possible extent, the bending of facts obtained through intelligence to suit a particular occupational viewpoint.

At the time of Pearl Harbor high officials here, despite warnings from our outstanding Ambassador to Japan, my old friend Joseph C. Grew, were convinced that the Japanese, if they struck, would strike southward against the soft underbelly of the British, French and Dutch colonial area. The likelihood that they would make the initial move against their most dangerous antagonist, the United States, was discounted. The attacks on Hawaii and the Philippines, and the mishandling of the intelligence we then had, greatly influenced our government's later decision on how our intelligence work should be organized. While the warnings received before the attack from deciphered Japanese cables may not have been clear enough to permit our leaders to pinpoint Hawaii and the Philippines, they should at least, if adequately analyzed, have alerted us to imminent danger in the Pacific.

If anyone has any doubt about the importance of objective intelligence, I would suggest a study of other mistakes which leaders have made because they were badly advised or misjudged the actions or reactions of other countries. When Kaiser Wilhelm II struck at France in 1914, he was persuaded by his military leaders that the violation of Belgian neutrality was essential to military success. He relied too heavily on their judgment and disregarded the advice he received from the political side as to the consequences of British intervention.

In the days prior to World War II, the British Government, despite Churchill's warnings, failed to grasp the dimensions of the Nazi threat, especially in aircraft.

Hitler likewise, as he launched into World War II, made a series of miscalculations. He discounted the strength and determination of Britain; later he opened a second front against Russia in June, 1941, with reckless disregard of the consequences. When in 1942 he was reportedly advised of the plan for an American-British landing in North Africa, he refused to pay attention to the intelligence available to him. I was told that he casually remarked, "They don't have the ships to do it."

As for Japan, successful as was the Pearl Harbor attack, later events proved that its government made the greatest miscalculation of all when it underestimated United States military potential.

Today a new threat, practically unknown in the days before the Communist revolution, has put an added strain on our intelligence capabilities. It is the Communist attempt—which we began to comprehend after World War II—to undermine the security of free countries. As this is carried on in secret, it requires secret intelligence techniques to ferret it out and to build up our defenses against it.

In the Soviet Union we are faced with an antagonist that has raised the art of espionage to an unprecedented height, while developing the collateral techniques of subversion and deception into a formidable political instrument of attack. No other country has ever before attempted this on such a scale. These operations, in support of the U.S.S.R.'s overall policies, go on in times of so-called thaw and under the guise of coexistence with the same vigor as in times of acute crisis. Our intelligence has a major share of the task of neutralizing such hostile activities, which present a common danger to us and to our allies.

The fact that so many Soviet cases of both espionage and subversion have been uncovered in recent times and in several NATO countries is not due to mere accident. It is well that the world should know what the Soviets know already—namely, that the free countries of the world have been developing highly sophisticated counterintelligence organizations and have been increasingly effective over the years in uncovering Soviet espionage. Naturally, with our NATO and other alliances, we have a direct interest in the internal security arrangements of other countries with which secrets may be shared. If a NATO document is filched by the

Communists from one of our allies, it is just as harmful to us as if it were stolen from our own files. This creates an important requirement for international cooperation in intelligence work.

Our allies, and many friendly countries which are not formal allies, generally share our view of the Communist threat. Many of them can make and are making real contributions to the total strength of the Free World, including one in the intelligence field, to help keep us forewarned. However, some of our friends do not have the resources to do all they might wish, and they look to the United States for leadership in the intelligence field, as in many others. As we uncover hostile Communist plans, they expect us to help them in recognizing the threats to their own security. It is in our interest to do so. One of the most gratifying features of recent work in intelligence, and one that is quite unique in its long history, has been the growing cooperation established between the American intelligence services and their counterparts throughout the Free World which make common cause with us as we face a common peril.

There is a fundamental question about our intelligence work which, I realize, worries a good many people. Is it necessary, they ask, for the United States, with its high ideals and its traditions to involve itself in espionage, to send U-2s over other people's territory, to break other people's coded messages?

Many people who understand that such activities may be necessary in wartime still doubt that they are justified in time of peace. Do we spy on friend and foe alike, and do we have to do it merely because another less scrupulous and less moral type of country does it to us? I do not consider such questions improper, frivolous or pacifist. Indeed, it does us credit that these questions are raised.

Personally, I see little excuse for peacetime spying on our friends or allies. Apart from the moral issues, we have other and far more important ways of using our limited intelligence resources. Also, there are other ways of getting the information we need through normal diplomatic channels. Of course, we have to take into account the historical fact that we have had friends who became enemies—Germany on two recent occasions, and Italy and Japan. Hence, it is always useful to have "in the bank" a store of basic intelligence—most of it not very secret—about all countries. I recall that in the early days of World War II a call went out to the public for personal photographs of various areas of the world, particularly the islands of the Pacific. We did not then have adequate knowledge of the beaches and the flora and fauna of many places where our forces might shortly be landing.

But the answer to the question of the need for intelligence, particularly on the Communist bloc, is that we are not really "at peace" with them, and we have not been since Communism declared its own war on our system of government and life. We are faced with a closed, conspiratorial, police-dominated society. We cannot hope to maintain our position securely if this opponent is confident that he can surprise us by attacking the Free World at the time and place of his own choosing and without any forewarning.

4

The Task of Collection

The collection of foreign intelligence is accomplished in a variety of ways, not all of them either mysterious or secret. This is particularly true of overt intelligence, which is information derived from newspapers, books, learned and technical publications, official reports of government proceedings, radio and television. Even a novel or a play may contain useful information about the state of a nation.

Two sources of overt intelligence in the Soviet Union are, of course, the newspapers *Izvestia* and *Pravda*, which translate into *News* and *Truth*. The former is an organ of the government and the latter of the party. There are also "little" *Izvestias* and *Pravdas* throughout Russia. A wit once suggested that in *Izvestia* there is no news and in *Pravda* there is no truth. This is a fairly accurate statement, but it is, nevertheless, of real interest to know what the Soviets publish and what they ignore, and what turn they give to embarrassing developments that they are obliged to publish.

It is, for example, illuminating to compare the published text of Khrushchev's extemporaneous remarks in Soviet media with what he actually said. His now-famous retort to Western diplomats at a Polish Embassy reception in Moscow on November 18, 1956, "We will bury you," was not quoted thus in the Soviet press reports, even though it was overheard by many. The state press apparently has the right to censor Premier Khrushchev, presumably with his approval. Later, however, what Khrushchev then said caught up with him and he gave a lengthy and somewhat mollifying interpretation of it. Consequently, how and why a story is twisted is at least as interesting as the actual content. Often there is one version for domestic consumption, another for the other Communist bloc countries and still other versions for different foreign countries. There are times when the "fairy stories" that Communist regimes tell their own people are indicative of new vulnerabilities and new fears.

The collection of overt foreign information by the United States is largely the business of the State Department, with other government departments cooperating in accordance with their own needs. The CIA has an interest in the "product" and shares in collection, selection and translation. Obviously, to collect and sort out such intelligence on a world-wide basis is a colossal task, but the work is well organized and the burden equitably shared. The monitoring of foreign radio broadcasts that might be of interest to us is one of the biggest parts of the job. In the Iron Curtain countries alone, millions of words are spewed out over the air every day; most of the broadcasts of real interest originate in Moscow and Peking, some directed to domestic audiences and others beamed abroad.

All overt information is grist for the intelligence mill. It is there for the getting, but large numbers of trained personnel are required to cull it in order to find the grain of wheat in the mountains of chaff. For example, in the fall of 1961 we were forewarned by a few hours of the Soviet intention to resume atomic testing by means of a vague news item transmitted by Radio Moscow for publication in a provincial Soviet journal. A young lady at a remote listening post spotted this item, analyzed it correctly and relayed it to Washington immediately. Her vigilance and perceptiveness succeeded in singling out one significant piece of intelligence from the torrents of deadly verbiage that have to be listened to daily.

In countries that are free, where the press is free and the publication of political and scientific information is not hampered by the government, the collection of overt intelligence is of particular value and is of direct use in the preparation of our intelligence estimates. Since we are that kind of country ourselves, we are subject to that kind of collection. The Soviets pick up some of their most valuable information about us from our publications, particularly from our technical and scientific journals, published transcripts of Congressional hearings and the like. For the collection of this kind of literature, they often make use of the personnel of the satellite diplomatic missions in Washington. There is no problem in acquiring it. The Soviets simply want to spare themselves the tasks; also, they feel that a Polish or Czech collection agent is likely to be less conspicuous than a Russian.

Information is also collected in the ordinary course of conducting official relations with a foreign power. This is not overt in the sense that it is available to anyone who reads the papers or listens to the radio. Indeed, the success of diplomatic negotiations calls for a certain measure of secrecy. But information derived from diplomatic exchanges is made available to the intelligence service for the preparation of estimates. Such information may contain facts, slants and hints that are significant, especially when coupled with intelligence from other sources. If the Foreign Minister of X hesitates to accept a United States offer on Monday, it may be that he is seeing the Soviets on Tuesday and hoping for a better offer there. Later, from an entirely different quarter, we may get a glimpse into the Soviet offer. Together these two items will probably have much more meaning than either would have had alone.

The effort of overt collection is broad and massive. It tries to miss nothing that is readily available and might be of use. Yet there may be some subjects on which the government urgently needs information that are not covered by such material. Or this material may lack sufficient detail, may be inconclusive or may not be completely trustworthy. Naturally, this is more often the case in a closed society. We cannot depend on the Soviets making public, either intentionally or inadvertently, what our government most wants to know; only what they wish us to believe. When they do give out official information, it cannot always be trusted. Published statistics may credit a five-year plan with great success; economic intelligence from inside informants may show that the plan failed in certain respects and that the ruble statistics given were not a true index of values. Photographs may be doctored, or even faked, as was the famous Soviet publicity picture of the junk heap first designated as the downed U-2. The rocket in the Red Army Day parade, witnessed and photographed by Western newsmen and military attachés, may be a dud, an assemblage of odd rocket parts that do not really constitute a working missile. Easy as it is to collect overt intelligence, it is equally easy to plant deception within it. For all these reasons clandestine intelligence collection (espionage) must remain an essential and basic activity of intelligence.

Clandestine intelligence collection is chiefly a matter of circumventing obstacles in order to reach an objective. Our side chooses the objective. The opponent has set up the obstacles. Usually he knows which objectives are most important to us, and he surrounds these with appropriately difficult obstacles. For example, when the Soviets started testing their missiles, they chose launching sites in their most remote and unapproachable wastelands. The more closed and rigid the control a government has over its people, the more obstacles it throws up. In our time this means that U.S. intelligence must delve for the intentions and capabilities of a nation pledged to secrecy and organized for deception, whose key military installations may be buried a thousand miles off the beaten track.

Clandestine collection uses people: "agents," "sources," "informants." It may also use machines, for there are machines today that can do things human beings cannot do and can "see" things they cannot see. Since the opponent would try to stop this effort if he could locate and reach it, it is carried out in secret; thus we speak of it as clandestine collection. The traditional word for it is "espionage."

The essence of espionage is access. Someone, or some device, has to get close enough to a thing, a place or a person to observe or discover the desired facts without arousing the attention of those who protect them. The information must then be delivered to the people who want it. It must move quickly or it may get "stale." And it must not get lost or be intercepted en route.

At its simplest, espionage is nothing more than a kind of well-concealed reconnaissance. This suffices when a brief look at the target is all that is needed. The agent makes his way to an objective, observes it, then comes back and reports what he saw. The target is usually fairly large and easily discernible—such things as troop dispositions, fortifications or airfields. Perhaps the agent can also make his way into a closed installation and have a look around, or even make off with documents. In any case, the length of his stay is limited. Continuous reportage is difficult to maintain when the agent's presence in the area is secret and illegal.

Behind the Iron Curtain today, this method of spying is hardly adequate—not because the obstacles are so formidable that they cannot be breached, but because the kind of man who is equipped by his training to breach them is not likely to have the technical knowledge that will enable him to make a useful report on the complex targets that exist nowadays. If you don't know anything about nuclear reactors, there is little you can discover about one, even when you are standing right next to it. And even for the rare person who might be technically competent, just getting close to such a target is hardly enough to fulfill today's intelligence requirements. What is needed is a thorough examination of the actual workings of the reactor. For this reason it is unrealistic to think that U.S. or other Western tourists in the Soviet Union can be of much use in intelligence collection. But for propaganda reasons, the Soviets continue to arrest tourists now and then in order to give the world the impression that U.S. espionage is a vast effort exploiting even the innocent traveler.

Of far more long-term value than reconnaissance is "penetration" by an agent, meaning that he somehow is able to get inside the target and stay there. One of the ways of going about this is for the agent to insinuate himself into the offices or the elite circles of another power by means of subterfuge. He is then in a position to elicit the desired information from persons who come to trust him and who are entirely unaware of his true role. In popular parlance, this operation is called a "plant," and it is one of the most ancient devices of espionage. The case of Ben Franklin's secretary, Edward Bancroft, which I related in an earlier chapter, is a classical example of the planted agent.

A penetration of this kind is predicated upon a show of outer loyalties, which are often not put to the test. Nor are they easily tested, especially when opponents share a common language and background. But today, when the lines that separate one nation and one ideology from another are so sharply drawn, the dissembling of loyalties is more difficult to maintain over a long period of time and under close scrutiny. It can be managed, though. One of the most notorious Soviet espionage operations before and during World War II was the network in the Far East, directed by Richard Sorge, a German who was working in Tokyo as a correspondent of the *Frankfurter Zeitung*. Sorge made it his business to cultivate his fellow countrymen at the German embassy in Tokyo, and eventually succeeded in having himself assigned to the embassy's Press Section. This not only gave him excellent cover for secret work with his Japanese agents, but also provided him directly with inside information about the Nazis' conduct of the war and their relations with Japan.

To achieve this, Sorge had to play the part of the good Nazi, which he apparently did convincingly even though he detested the Nazis. The Gestapo chief in the embassy, as well as the ambassador, and the service attachés were all his "friends." Had the Gestapo in Berlin ever investigated Sorge's past, as it eventually did after Sorge was apprehended by the Japanese in 1941, it would have discovered that Sorge had been a Communist agent and agitator in Germany during the early 1920s and had spent years in Moscow.

Shortly thereafter, the West was subjected to similar treatment at the hands of Soviet espionage. Names such as Bruno Pontecorvo and Klaus Fuchs come to mind as agents who were unmasked after the war. In some such cases, records of previous Communist affiliations lay in the files of Western security and intelligence services even while the agents held responsible positions in the West, but they were not found until it was too late. Because physicists like Fuchs and Pontecorvo moved from job to job among the Allied countries—one year in Great Britain, another in Canada and another in the United States—and because the scientific laboratories of the Allies were working under great pressures, personnel with credentials from one Allied country were sometimes accepted for employment in another under the impression that they had already been sufficiently checked out. And when available records were consulted, the data found in them—particularly if of Nazi origin—seem often to have been discounted at a time when Russia was our ally and Hitler our enemy, and when the war effort required the technical services of gifted scientists of many nationalities.

The consequences of these omissions and oversights during the turbulent war years are regrettable, and the lesson will not easily be forgotten. We cannot afford any more Fuchses or Pontecorvos. Today investigation of persons seeking employment in sensitive areas of the U.S. Government and related technical installations is justifiably thorough and painstaking.

Consequently, an agent who performs as a plant in our time must have more in his favor than acting ability. With our modern methods of security checking, he is in danger of failure if there is any record of his ever having been something other than what he represents himself to be. The only way to disguise a man today so that he will be acceptable in hostile circles for any length of time is to make him over entirely. This involves years of training and a thorough concealing and burying of the past under layers of fictitious personal history which have to be "backstopped."

If you were really born in Finland but are supposed to have been born in Munich, Germany, then you must have documents showing your connection to that city. You have to be able to act like someone who was born and lived there. Arrangements have to be made in Munich to confirm your origin in case an investigation is ever undertaken. Perhaps Munich or a similar city was chosen because it was bombed and certain records were destroyed. A man so made over is known as an "illegal," and I shall have more to say about him later. Obviously, an intelligence service will go to all this trouble only when it is intent upon creating deep-set and long-range assets.

If an intelligence service cannot insert its own agent within a highly sensitive target, the alternative is to recruit somebody who is already there. You might find someone who is inside but is not quite at the right spot for access to the information you need. Or you might find someone just beginning a career which will eventually lead to his employment in the target. But the main thing is that he is a qualified and "cleared" insider. He is, as we say, "in place."

One of my most valuable agents during World War II, of whom I shall have more to say later, was precisely of this kind. When I first established contact with him, he was already employed in the German Foreign Office in a position which gave him access to communications with German diplomatic establishments all over the world. He was exactly at the right place. No single diplomat abroad, of whatever rank, could have got his hands on so much information as did this man, who had access to the all-important Foreign Office files. Even with the most careful planning many years in advance, it would have been a stroke of fortune if we could ever have placed an agent inside this target and maneuvered him into such a position, even if he had been able to behave like the most loyal Nazi. This method of recruiting the agent "in place," despite its immense difficulties, has the advantage of allowing the intelligence service to focus on the installation it wishes to penetrate, to examine and analyze it for its most important and most vulnerable points, and then to search for the man already employed at that point who might be likely to cooperate. It does not, as in the case of plants, begin with the man, the agent, and hope it can devise a way of inserting him into the target.

In recent years, most of the notorious instances of Soviet penetration of important targets in Western countries were engineered in this way, by the recruitment of someone already employed inside the target.

David Greenglass at Los Alamos during World War II, though only a draftsman, had access to secret details of the internal construction of the atomic bomb. Judith Coplon was employed shortly after the war in a section of the Department of Justice responsible for the registration of foreign agents in the United States. She regularly saw and copied for the Soviets FBI reports which came across her desk on investigations of espionage in the United States. Harry Houghton and John Vassall, although of low rank and engaged chiefly in administrative work, were able to procure sensitive technical documents from the British Admiralty, where they were employed in the late 1950s. Alfred Frenzel, a West German parliamentarian, had access to the NATO documents which were distributed to a West Germany Parliamentary Defense Committee on which he served in the mid-1950s. Irvin Scarbeck was only an administrative officer in our embassy in Warsaw in 1960–61. But after he had been compromised by a Polish girl and blackmailed, he managed to procure for the Polish Intelligence Service, which was operating under Soviet direction, some of our ambassador's secret reports to the State Department on the political situation in Eastern Europe.

All these people were already employed in jobs which made them interesting to the Communists at the time they were first recruited. Some of them moved up later into jobs which made them of even greater value to the Soviets. In some instances this may have been achieved with secret Soviet guidance. Houghton and Vassall were both originally recruited while stationed at British embassies behind the Iron Curtain. When each was returned home and assigned to a position in the Admiralty, his access to important documents naturally broadened. Similarly, had Scarbeck not been caught as a result of careful counterintelligence efforts while still at his post in Warsaw, he probably could have continued for years to be of ever-increasing use to the Soviets as he was reassigned to one United States diplomatic post after another.

The Soviet Union gave widespread publicity to the case of an "insider" who worked with Western intelligence and who they admitted had access to information of great value. This was the case of Colonel Oleg Penkovsky, whose conviction and execution by the Soviets are now a matter of history. His trial, along with that of the Englishman Greville Wynne, lasted just one week in early May of 1963. It is not entirely clear just why the Soviets chose to make a "show trial" of this case rather than to keep the whole affair entirely secret, which it was certainly in their power to do. The most likely reason was to discourage further espionage among their own people by showing them that in the end the culprit always gets caught. This, of course, is not true. But in staging the trial, they openly admitted that Penkovsky had caused them very considerable damage.

It is fairly plain from the evidence which the Soviets allowed to be presented in the court that a combination of Western intelligence services had succeeded a few years back in gaining the services of the Soviet colonel, who held an important position in the military and technical hierarchy of the Red Army. Penkovsky was trusted by the Soviets and allowed to travel to various international conferences in Western Europe. These afforded the occasions for establishing contact and communication with Penkovsky.

The Soviets claim that he was lured by material attractions—wine, women and song—available in the West. This is the usual method of discrediting an individual whose actions and motives may, in fact, have been far worthier than they are willing to admit. But Penkovsky was a high-level and experienced officer with many high Soviet decorations and not some youthful adventurer, not a man likely to fall for material benefits alone. There must have been much more involved than the trial and publicity indicate. The Soviet hierarchy has been deeply shaken, for Penkovsky had lost faith in the system that employed him.

Whatever his motives, the case is typical of the current pattern of espionage. Penkovsky had natural access to important information. All his advantages were built in. No reconnaissance, no traveler, no plant could have duplicated his achievement. He was already there. He had to be discovered, contact had to be established with him, he had to be convinced that he could make a valuable contribution to a cause in which he believed.

A similar case, which also ended tragically for the agent, was that of the Bulgarian diplomat Asen Georgieff, who was tried and executed in Sofia for espionage in December, 1963. During his trial, there was a great deal of propaganda given out by the Bulgarians concerning Georgieff's alleged weakness for the material benefits of the West. Little was said about the fact that Georgieff had long been a Communist intellectual of unusually high caliber, a doctor of laws, an internationally recognized Hegel scholar, a man whose mental prowess placed him head and shoulders above his colleagues and had earned for him one of the top-ranking positions in his country's delegation to the Untied Nations. He was not, as were most of his colleagues, chosen for this position because of party accomplishments.

Unlike Penkovsky, whose contributions were in the field of military and technical intelligence, Georgieff, according to indications which came out during his trial, was of interest to Western intelligence because of his access to political information. East and West guard their major military and technical secrets with about equal fervor, if not always equal success. On the other hand, much of "political intelligence" is no secret at all in the West, but is regarded as highly sensitive information in the Soviet-satellite areas. The U.S. Congress debates openly, and the results of the deliberations of the cabinet and even of the National Security Council sooner or later tend to reach the public. The equivalent deliberations of the Kremlin and of the politburos of the satellites are matters of deepest secrecy, thus necessitating an intelligence effort to uncover them.

The overt and clandestine methods of collection I have been discussing are obviously quite inadequate alone to meet all our intelligence needs today. They can be and are supplemented by other methods, particularly by taking advantage of the great advances in science and technology and through the fact that much intelligence comes to us from "volunteers," about whom I shall have much to say later.

5

Collection—Enter the Machine

The intelligence service needs a man who speaks Swahili and French, has a degree in chemical engineering, is unmarried and over thirty-five but under five feet eight. You push a button and in less than forty seconds a machine—like those commonly used in personnel work—tells whether such a man is available, and if so, everything else there is on record about him. Similar machines are used in sorting and assembling the data of intelligence itself.

This means that among the ranks of the analysts and evaluators in intelligence work today there are also persons trained in data processing and in the handling of computers and other complex "thinking" machines.

We are under no illusions that these machines improve the nature of the information. This will always depend on the reliability of the source and the skill of the analyst. What machines can do, however, is recover quickly and accurately from the enormous storehouse of accumulated information such past data as are necessary for evaluating current information. What, before the advent of the machine, might have taken the analyst weeks of search and study among the files, the machines can now accomplish in a matter of minutes.

But this is an ordinary feat compared to what technology can do today in collecting the information itself. Here I am speaking not of computers and business machines, but of special devices which have been developed to observe and record events, to replace in a sense the human hand and eye or to take over in areas which human capabilities cannot reach.

The technical nature of many contemporary targets of intelligence has itself suggested or prompted the creation of the devices which can observe them. If a target emits a telltale sound, then a sensitive acoustical device comes to mind for monitoring and observing it. If the target causes shock waves in the earth, then seismographic apparatus will detect it.

Moreover, the need to observe and measure the effects of our own experiments with nuclear weapons and missiles hastened the refinement of equipment which, with some modifications, can also be useful for watching other people's experiments. Radar and accurate long-range photography are basic tools of technical collection. Another is the collection and analysis of air samples in order to determine the presence of radioactivity in the atmosphere. Since radioactive particles are carried by winds over national borders, it is unnecessary to penetrate the opponent's territory by air or land in order to collect such samples.

In 1948 our government instituted round-the-clock monitoring of the atmosphere by aircraft for detecting experimentation with atomic weapons. The first evidence of a Soviet atomic explosion on the Asiatic mainland was detected by this means in September of 1949, to the surprise of the world and of many scientists who until then had believed, on the basis of available evidence, that the Soviets would not "have the bomb" for years to come. Refinements in instrumentation then began to reveal to us not only the fact that atomic explosions had taken place but also the power and type of the device or weapon detonated.

Such developments, as was to be expected, eventually inspired the opponent, who learned that his experiments were being monitored, to take countermeasures, also of a highly technological nature. It is now possible to "shield" atomic explosions both underground and in the outer atmosphere so that their characteristics cannot be easily identified as to size and type. The next round, of course, is for the enterprising technicians on the collection side to devise means of penetrating the countermeasures.

The protracted negotiations with the Soviets in recent years on the subject of disarmament and the nuclear test ban involve precisely these problems and have brought out into the open the amazingly complex research, hitherto secret, which we and the Soviets also are devoting to the problems both of shielding experiments with nuclear devices and of detecting them even when they are shielded.

Modern technology thus tries to monitor and observe certain scientific and military experiments of other nations by concentrating on the "side effects" of their experiments. Space research presents quite another kind of opportunity for monitoring. Space vehicles while in flight report back data on their performance as well as on conditions in outer space or in the neighborhood of heavenly bodies by means of electronic signals, or telemetry. These signals are of course meant for the bases and stations of the country that sent the vehicle aloft. Since, as in the case of ordinary radio messages, there is nothing to stop anyone with the right equipment from "listening in," it is obvious that nations competing in space experimentation are going to intercept each other's telemetry in an attempt to find out what the other fellow's experiments are all about and how well they have succeeded. The trick is to read the signals right.

Many important military and technical targets are, however, static and do not betray their location or the nature of their activity in ways which can be detected, tracked, monitored or intercepted. Factories, shipyards, arsenals, missile bases under construction do not give off telltale evidence of their existence which can be traced from afar. To discover the existence of such installations one must get close to them or directly over them at very high altitudes, armed with long-range cameras. This was, of course, the purpose of the U-2, which could collect information with more speed, accuracy and dependability than could any agent on the ground. In a sense, its feats could be equaled only by the acquisition of technical documents directly from Soviet offices and laboratories. The U-2 marked a new high, in more ways than one, in the scientific collection of intelligence. Thomas S. Gates, Jr., Secretary of Defense of the United States at the time of the U-2 incident, May 1, 1960, testified to this before the Senate Foreign Relations Committee on June 2, 1960:

> From these flights we got information on airfields, aircraft, missiles, missile testing and training, special weapons storage, submarine production, atomic production and aircraft deployment . . . all types of vital information. These results were considered in formulating our military programs. We obviously were the prime customer, and ours is the major interest.

In more recent days, it was the high-altitude U-2 reconnaissance flights which gave the "hard" evidence of the positioning in Cuba of Soviet medium-range missiles in late October of 1962. If they had not been discovered while work on the bases was still in progress and before they could be camouflaged, these bases might have constituted a secret and deadly threat to our security and that of this hemisphere. Here, too, was an interesting case in which classical collection methods wedded to scientific methods brought extremely valuable results. Various agents and refugees from Cuba reported that something in the nature of missile bases was being constructed and pinpointed the area of construction; this led to the gathering of proof by aerial reconnaissance.

The question whether the piloted U-2 can be superseded by pilotless satellites orbiting the globe at much higher altitudes came up in May, 1964, when Premier Khrushchev declared that the United States could avoid international tension by desisting from further flights of the U-2 over Cuba. The space satellites, said Khrushchev, can do the same job, and he offered to show our President photographs of American military bases taken by Soviet "sky spies." I doubt whether we would agree wholly with Khrushchev that space vehicles should supersede the manned plane for all reconnaissance purposes. But his admissions of the use to which his satellites have been put is an interesting one.

Eloquent testimony to the value of scientific intelligence collection, which has proved its worth a hundred times over, has been given by Winston Churchill in his history of World War II.[1] He describes British use of radar in the Battle of Britain in September, 1940, and also tells of bending, amplifying and falsifying the direction signals sent by Berlin to guide the attacking German aircraft. Churchill calls it all the "wizard war" and he concludes that "Unless British science had proved superior to German and unless its strange, sinister resources had been effectively brought to bear in the struggle for survival, we might well have been defeated, and being defeated, destroyed."

Science as a vital arm of intelligence is here to stay. We are in a critical competitive race with the scientific development of the Communist bloc, particularly that of the Soviet Union, and we must see to it that we *remain* in a position of leadership. Some day this may be as vital to us as radar was to Britain in 1940.

[1]*The Second World War* (Boston: Houghton, Mifflin Co., 1948–53).

AUDIO SURVEILLANCE

A technical aid to espionage of another kind is the concealed microphone and transmitter which keeps up a flow of live information from
inside a target to a nearby listening post; this is known to the public as
"telephone tapping" or "bugging" or "miking." "Audio surveillance,"
as it is called in intelligence work, requires excellent miniaturized electronic equipment, clever methods of concealment and a human agent to
penetrate the premises and do the concealing.

Ambassador Henry Cabot Lodge in early June of 1960 displayed
before the United Nations in New York the Great Seal of the United
States which had been hanging in the office of the American Ambassador in Moscow. In it the Soviets had concealed a tiny instrument
which, when activated, transmitted to a Soviet listening post everything
that was said in the Ambassador's office. Actually, the installation of
this device was no great feat for the Soviets since every foreign embassy
in Moscow has to call on the services of local electricians, telephone
men, plumbers, charwomen and the like. The Soviets have no difficulties in seeing to it that their own citizens cooperate with their intelligence service, or they may send intelligence officers, disguised as
technicians, to do the job.

In early May, 1964, our State Department publicly disclosed that
as a result of a thorough demolishing of the internal walls, ceilings and
floors of "sensitive" rooms in our embassy in Moscow, forty concealed
microphones were brought to light. Previous intensive electronic testing for such hidden devices had not located any of these microphones.

In Soviet Russia and in the major cities of the satellite countries
certain hotel rooms are designated for foreign travelers because they
have been previously bugged on a permanent basis. Microphones do
not have to be installed in a rush when an "interesting" foreigner ar-
rives on the scene. The microphones are already there, and it is only
the foreigner who has to be installed. All the hotels are state-owned
and have permanent police agents on their staffs whose responsibility
is to see that the proper foreigners are put in the "right" rooms.

When Chancellor Adenauer paid his famous visit to Moscow in September, 1955, to discuss the resumption of diplomatic relations between
Russia and West Germany, he traveled in an official German train. When
he arrived in Moscow, the Soviets learned to their chagrin that the wily
Chancellor (who then had no embassy of his own to reside in, for such

limited security as this might afford) intended to live in his train during his stay in Moscow and did not mean to accept Soviet "hospitality" in the form of a suite at one of the VIP hotels for foreigners in Moscow. It is reported that before leaving Germany the Chancellor's train had been equipped by German technicians with the latest devices against audio surveillance.

Outside its own country an intelligence service must consider the possible repercussions and embarrassments that may result from the discovery that an official installation has been illegally entered and its equipment tampered with. As in all espionage operations, the trick is to find the man who can do the job and who has the talent and the motive, whether patriotic or pecuniary. There was one instance when the Soviets managed to place microphones in the flowerpots that decorated the offices of a Western embassy in a neutral country. The janitor of the building, who had a weakness for alcohol, was glad to comply for a little pocket money. He never knew who the people were who borrowed the pots from him every now and then or what they did with them.

There is hardly a technological device of this kind against which countermeasures cannot be taken. Not only can the devices themselves be detected and neutralized, but sometimes they can be turned against those who install them. Once they have been detected, it is often profitable to leave them in place in order to feed the other side with false or misleading information.

In their own diplomatic installations abroad, the Soviets and their satellites stand in such fear of audio surveillance operations being mounted against them that they will usually refuse to permit local service people to install telephones or even ordinary electrical wiring in buildings they occupy. Instead, they will send out their own technicians and electricians as diplomats on temporary duty and will have them do the installing. In one instance where they evidently suspected that one of their embassies had been "wired for sound" by outsiders, they even sent a team of day laborers to the capital in questions, all of them provided with diplomatic passports for the trip. To the great amusement of the local authorities, these "diplomats" were observed during the next few weeks in overalls and bearing shovels, digging a trench four or five feet deep in the ground around the embassy building, searching for buried wires leading out of the building. (They didn't find any.)

CODES AND CIPHERS

"Gentlemen," said Secretary of State Stimson in 1929, "do not read each other's mail," and so saying, he shut down the only American cryptanalytic (code-breaking) effort functioning at that time. Later, during World War II, when he was serving as Secretary of War under President Franklin D. Roosevelt, he came to recognize the overriding importance of intelligence, including what we now call "communications intelligence." When the fate of a nation and the lives of its soldiers are at stake, gentlemen do reach each other's mail—if they can get their hands on it.

I am, of course, not speaking here of ordinary mail, although postal censorship has itself often played a significant role in intelligence work. However, except in the detection of secret writing, there is little technology involved in postal censorship. Modern communications intelligence, on the other hand, is a highly technical field, one that has engaged the best mathematical minds in an unceasing war of wits that can easily be likened to the battle for scientific information which I described a little earlier.

Every government takes infinite pains to invent unbreakable systems of communication and to protect these systems and the personnel needed to run them. At the same time, it will do everything in its power to gain access or insight into the communications of other governments whose policies or actions may be of real concern to it. The reasons for this state of affairs on both sides is obvious. The contents of official government messages, political or military, on "sensitive" subjects constitute, especially in times of crisis, the best and "hottest" intelligence that one government can hope to gather about another.

There is a vast difference between the amateur and professional terminology in this field. If I stick to the amateur terms, I shall probably offend the professionals, and if I use the professional terms, I shall probably bore and confuse the amateur. My choice is an unhappy one and I will be brief. In a code, some word, symbol or group of symbols is substituted for a whole word or even for a group of words or a complete thought. Thus, "XLMDP" or "79648," depending upon whether a letter or number code is used, could stand for "war" and every time they turn up in a message that is what they mean. When the Japanese Government set up the famous "East Winds" code for their diplomats in the United States in December, 1941, they were prepared to indicate through the simplest prearranged code words that an attack in the Pacific was forthcoming.

In a cipher, a symbol, such as a letter or number, stands for a single letter in a word. Thus, "b" or "2" can mean "e" or some other letter. In simple ciphers the same symbol always stands for the same letter. In the complex ciphers used today, the same symbol can stand for a different letter each time it turns up. Sometimes a message is first put into code, and then the code is put into cipher.

The United States military forces were able to resort to rather unusual "ready-made" codes during World War I, and in a few instances during World War II, in communications between units in the field. These resources were our native American Indian languages, chiefly the Navajo language, which has no written forms and had never been closely studied by foreign scholars. Two members of the same tribe at either end of a field telephone could transmit messages which no listener except another Navajo could possibly understand. Needless to say, neither the Germans nor the Japanese had any Navajos.

In modern terminology, the word "crypt," meaning "something hidden," conveniently gets around the distinction between codes and ciphers since it refers to all methods of transforming "plain text" or "clear text" into symbols. The over-all term for the whole field today is "cryptology." Under this broad heading we have two distinct areas. Cryptography has to do with making, devising, inventing or protecting codes and ciphers for the use of one's own government. Cryptanalysis, on the other hand, has to do with breaking codes and ciphers or "decrypting" them, with translating someone else's intercepted messages into proper language. To put one's own messages into a code or cipher is to "encrypt" them. However, when we translate our own messages back into plain language, we are "deciphering."

A cryptogram or cryptograph would be any message in code or cipher. "Communications intelligence" is information which has been gained through successful cryptanalysis of other people's traffic. And now, having confused the reader completely, we can get to the gist of the matter.

The diplomatic service, the armed services and the intelligence service of every country use secret codes and ciphers for classified and urgent long-distance communications. Transmission may be via commercial cable or radio or over special circuits set up by governments. Anyone can listen in to radio traffic. Also, governments, at least in times of crisis, can usually get copies of the encrypted messages that foreign diplomats stationed on their territory send home via commercial cable facilities. The problem is to break the codes and ciphers, to "decrypt" them.

Certain codes and ciphers can be broken by mathematical analysis of intercepted traffic, i.e., cryptanalysis, or more dramatically and simply by obtaining copies of codes or code books or information on cipher machines being used by an opponent, or by a combination of these methods.

In the earlier days of our diplomatic service, up to World War I, the matter of codes was sometimes treated more or less cavalierly, often with unfortunate results. I remember a story told me as a warning lesson when I was a young foreign service officer. In the quiet days of 1913, we had as our Minister in Rumania an estimable politician who had served his party well in the Midwest. His reward was to be sent as Minister to Bucharest. He was new to the game and codes and ciphers meant little to him. At that time our basic system was based on a book code, which I will call the Pink Code, although that was not the color we then chose for its name. I spent thousands of worried hours over this book, which I have not seen for over forty years, but to this day I can still remember that we had six or seven words for "period." One was "PIVIR" and another was "NINUD." The other four or five I do not recall. The theory than was—and it was a naïve one—that if we had six or seven words it would confuse the enemy as to where we began and ended our sentences.

In any event, our Minister to Rumania started off from Washington with the Pink Code in a great, sealed envelope and it safely reached Bucharest. It was supposed to be lodged in the legation's one safe. However, handling safe combinations was not the new Minister's forte, and he soon found it more convenient to put the code under his mattress, where it rested happily for some months. One day it disappeared—the whole code book and the Minister's only code book. It is believed that it found its way to Petrograd.

The new Minister was in a great quandary, which, as a politician, he solved with considerable ingenuity. The coded cable traffic to Bucharest in those days was relatively light and mostly concerned the question of immigrants to the United States from Rumania and Bessarabia. So when the new Minister had collected a half-dozen coded messages, he would get on the train to Vienna, where he would quickly visit our Ambassador. In the course of conversation, the visitor from Bucharest would casually remark that just as he was leaving he had received some messages which he had not had time to decode and could he borrow the Ambassador's Pink Code. (In those good old days, we sent the same code books to almost all of our diplomatic missions.) The Minister to Bucharest would then decipher his messages, prepare and code appropriate replies, take the train back to Bucharest and, at staged intervals, send off the coded replies. For a time everything went smoothly. The secret of the loss of the code book was protected until August, 1914, brought a flood of messages from Washington as the dramatic events leading up to World War I unrolled. The Minister's predicament was tragic—trips to Vienna no longer sufficed. He admitted his dereliction and returned to American politics.

The uncontrollable accidents and disasters of war sometimes expose to one opponent cryptographic materials used by the other. A headquarters or an outpost may be overrun and in the heat of retreat code books left behind. Many notable instances of this kind in World War I gave the British a lifesaving insight into the military and diplomatic intentions of the Germans. Early in the war the Russians sank the German cruiser *Magdeburg* and rescued from the arms of a drowned sailor the German naval code book, which was promptly turned over to their British allies. British salvage operations on sunken German submarines turned up similar findings. In 1917 two German dirigibles, returning from a raid over England, ran into a storm and were downed over France. Among the materials retrieved from them were coded maps and code books used by German U-boats in the Atlantic.

An American naval exploit which took place toward the end of World War II has given us an even more thrilling story of the capture of enemy code and cipher material. This was the result of a carefully laid plan and not of a lucky accident. A German submarine, the U-505, was captured, intact, on June 4, 1944, off the coast of French West Africa by units of the United States Navy under the command of Rear Admiral Daniel V. Gallery.

During World War II, Allied action resulted in the destruction of over seven hundred German U-boats. The U-505, which now reposes in the Museum of Science and Industry, Chicago, was the only one that was brought back afloat and in one piece. It had been the consistent practice of the German U-boat crews whose subs were forced to surface and surrender to insure that the submarine would sink as the crew abandoned ship. In this instance, however, as the result of skillful preparation, a boarding party from Admiral Gallery's task force managed to get abroad the U-505 just as its own crew was abandoning it after having set its valves for scuttling. At the risk of their lives and not knowing how many seconds they had before the submarine would take its final plunge, some ten men from the American naval boarding crew charged down the hatch and closed the scuttling valves just in the nick of time. Their escape was later aided by a German sailor. He had jumped overboard and was swimming near the sinking German sub when a member of the boarding crew hauled him aboard again and got him to disclose the workings of a conning tower hatch which was on the escape route of the Americans who had gone below. As they threw him back into the water, it was with a heartfelt "Thanks, bud," but rescue was at hand for him and the other German crew members.

All the records and files and technical equipment aboard the sub, including its codes and ciphers, were rescued, and the submarine was safely towed to Bermuda.

But this was not the end of the story. If the Nazis had learned that the submarine had not been scuttled or destroyed before capture, they would have been alerted to the probable seizure of the code and cipher material aboard and would never again have used them. Obviously several thousand American naval personnel, from the beginning to the end of the operation of capture and of towing, knew the facts, and for many this was their great story of the war. The problem of impressing upon all these sailors the importance of keeping the capture secret was a bigger task even than capturing the submarine itself. But this was done with success. The Germans believed that the submarine had gone to its watery grave, carrying with it the secrets which in fact proved very useful to us.[2]

Military operations based on breaking of codes will often tip off the enemy, however. When, during World War I, the Germans noticed that their submarines were being cornered with startling frequency, it was not hard for them to guess that communications with their underwater fleet were being read. As a result, all codes were immediately changed. There is always the problem, then, of how to act on information derived in this manner. One can risk terminating the usefulness of the source in order to obtain an immediate military or diplomatic gain, or one can hold back and continue to accumulate an ever-broadening knowledge of the enemy's movements and actions in order eventually to inflict the greatest possible damage.

[2]An account of this naval exploit appears in Daniel V. Gallery, *Twenty Billion Tons Under the Sea* (Chicago: Henry Regnery Co., 1954).

Actually, in either case, the attempt is usually made to protect the real source and keep it viable, by giving the enemy fake indications that some other kind of source was responsible for the information acquired. Sometimes an operation that could damage the adversary is not undertaken if it would alert the enemy to the fact that its origin was solely due to information obtained by reading his messages.

During World War I, the first serious American cryptanalytic undertaking was launched under the aegis of the War Department. Officially known as Section 8 of Military Intelligence, it liked to call itself the "Black Chamber," the name used for centuries by the secret organs of postal censorship of the major European nations. Working from scratch, a group of brilliant amateurs under the direction of Herbert Yardley, a former telegraph operator, had by 1918 become a first-rate professional outfit. One of its outstanding achievements after World War I was the breaking of the Japanese diplomatic codes. During negotiations at the Washington Disarmament Conference in 1921, the United States wanted very much to get Japanese agreement to a 10:6 naval ratio. The Japanese came to the conference with the stated intention of holding to a 10:7 ratio. In diplomacy, as in any kind of bargaining, you are at a tremendous advantage if you know your opponent is prepared to retreat to secondary positions if necessary. Decipherment of the Japanese diplomatic traffic between Washington and Tokyo by the Black Chamber revealed to our government that the Japanese were actually ready to back down to the desired ratio if we forced the issue. So we were able to force it without risking a breakup of the conference over the issue.

The "Black Chamber" remained intact, serving chiefly the State Department, until 1929, when Secretary Stimson refused to let the department avail itself further of its services. McGeorge Bundy, Stimson's biographer, provides this explanation:

> Stimson adopted as his guide in foreign policy a principle he always tried to follow in personal relations—the principle that the way to make men trustworthy is to trust them. In this spirit he made one decision for which he was later severely criticized: he closed down the so-called Black Chamber. . . . This act he never regretted . . . . Stimson, as Secretary of State, was dealing as a gentleman with the gentlemen sent as ambassadors and ministers from friendly nations.[3]

[3]Henry L. Stimson and McGeorge Bundy, *On Active Service in Peace and War* (New York: Harper & Brothers, 1948).

Our Army and Navy had, fortunately, continued to address themselves to the problems of cryptanalysis with particular emphasis on Japan, since American military thinking at that time foresaw Japan as the major potential foe of the United States in whatever war was to come next. By 1941, the year of Pearl Harbor, our cryptanalysts had broken most of the important Japanese naval and diplomatic codes and ciphers; and we were, as a result, frequently in possession of evidence of imminent Japanese action in the Pacific before it took place.

The Battle of Midway in June, 1942, the turning point of the naval war in the Pacific, was an engagement we sought because we were able to learn from decrypted messages that a major task force of the Imperial Japanese Navy was gathering off Midway. This intelligence concerning strength and disposition of enemy forces gave our Navy the advantage of surprise.

A special problem, in the years following Pearl Harbor, was how to keep secret the fact that we had broken the Japanese codes. Investigations, recriminations, the need to place the blame somewhere for the disheartening American losses threatened to throw this "Magic," as it was called, into the lap of the public, and the Japanese. Until an adequate Navy could be put on the seas, the ability to read Japanese messages was one of the few advantages we had in the battle with Japan. There were occasional leaks but none evidently ever came to their attention.

In 1944, Thomas E. Dewey, who was then running for President against President Roosevelt, had learned, as had many persons close to the federal government, about our successes with the Japanese code and our apparent failure before Pearl Harbor to make the best use of the information in our hands. It was feared that he might refer to this in his campaign. The mere possibility sent shivers down the spines of our Joint Chiefs of Staff. General Marshall himself then wrote a personal letter to Mr. Dewey, telling him that the Japanese still did not know we had broken their codes and that we were achieving military successes as a result of our interception and decoding of their messages. Mr. Dewey never mentioned our code successes. The secret was kept.

One of the most spectacular of all coups in the field of communications intelligence was the British decipherment of the so-called Zimmermann telegram in January, 1917, when the United States was on the brink of World War I.[4] The job was performed by the experts of "Room 40," as British naval cryptanalytic headquarters were called. The message had originated with the German Foreign Secretary Zimmermann in Berlin and was addressed to the German Minister in Mexico City. It outlined the German plan for the resumption of unrestricted submarine warfare on February 1, 1917, stated the probability that this would bring the United States into war, and proposed that Mexico enter the war on Germany's side and with victory regain its "lost territory in Texas, New Mexico, and Arizona."

Admiral Hall, the legendary Chief of British Naval Intelligence, had this message in his hands for over a month after its receipt. His problem was how to pass its decrypted contents to the Americans in a manner that would convince them of its authenticity yet would prevent the Germans from learning the British had broken their codes. Finally, the war situation caused Lord Balfour, the British Foreign Secretary, to communicate the Zimmermann message formally to the American Ambassador in London. The receipt of the message in Washington caused a sensation at the White House and State Department, and created serious problems for our government—how to verify beyond a doubt the validity of the message and how to make it public without letting it seem merely an Anglo-American ploy to get the United States into the War. My uncle, Robert Lansing, who was then Secretary of State, later told me about the dramatic events of the next few days which brought America close to war.

The situation was complicated by the fact that the Germans had used American diplomatic cable facilities to transmit the message to their Ambassador in Washington, Count Bernstorff. He relayed it to his colleague in Mexico City. President Wilson had granted the Germans the privilege of utilizing our communication lines between Europe and America on the understanding that the messages would be related to peace proposals in which Wilson was interested.

The President's chagrin was therefore all the greater when he discovered to what end the Germans had been exploiting his good offices. However, this curious arrangement turned out to be of great advantage. First of all, it meant that the State Department had in its possession a copy of the encrypted Zimmermann telegram, which it had passed to Bernstorff, unaware, of course, of its inflammatory contents. Once the

[4]This story has been well told in Barbara Tuchman's book. See Bibliography.

encrypted text was identified, it was forwarded to our embassy in London, where one of Admiral Hall's men redecrypted it for us in the presence of an embassy representative, thus verifying beyond a doubt its true contents. Secondly, the fact that deciphered copies of the telegram had been seen by German diplomats in both Washington and Mexico City helped significantly to solve the all-important problem that had caused Admiral Hall so much worry, namely, how to fool the Germans about the real source from which we had obtained the information. In the end the impression given the Germans was that the message had leaked as a result of some carelessness or theft in one of the German embassies or Mexican offices which had received copies of it. They continued using the same codes, thus displaying a remarkable but welcome lack of imagination. On March 1, 1917, the State Department released the contents of the telegram through the Associated Press. It hit the American public like a bombshell. In April we declared war on Germany.

When one compares the cryptographic systems used today with those to which governments during World War I entrusted the passage of their most vital and sensitive secrets, the latter seem crude and amateurish, especially because of their recurring groups of symbols which tipped off the cryptanalyst that an important word or one in frequent usage must lie behind the symbols. When Admiral Hall's cryptanalysts saw the combination "67893" in the Zimmermann telegram, they recognized it and knew that it meant "Mexico." Under the German system it always meant that. Today such a cipher group would never stand for the same word twice.

Today not only all official government messages but also the communications of espionage agents are cast in equally secure and complex cryptographic systems. Soviet agents, for example, in reporting information back to Moscow, use highly sophisticated cipher systems. Here as elsewhere, as defensive measures improve, countermeasures to pierce the new defenses also improve.

6

Planning and Guidance

The matters that interest an intelligence service are so numerous and diverse that some order must be established in the process of collecting information. This is logically the responsibility of the intelligence headquarters. It alone has the world picture and knows what the requirements of our government are from day to day and month to month.

Without guidance and direction, intelligence officers in different parts of the world could easily spend much of their time duplicating each other's work or there could be serious gaps in our information. The intelligence officer at his post abroad cannot fully judge the value of his own operations because he cannot know whether the information he is procuring has already been picked up somewhere else, or is known from overt sources, or is of too low a priority to be worth the effort or the expense.

Our government determines what the intelligence objectives are and what information it needs, without regard to obstacles. It also establishes priorities among these objectives according to their relative urgency. Soviet ICBMs will take priority over their steel production. Whether or not Communist China would to go war over Laos will take priority over the political shading of a new regime in the Middle East. Only after priority has been established is the question of obstacles examined. If the information can be obtained by overt collection or in the ordinary course of diplomatic work, the intelligence service will not be asked to devote to the task its limited assets for clandestine collection.

But if it is decided that secret intelligence must do the job, then it is usually because serious obstacles are known to surround the target.

In preparing its directives for the intelligence mission in a particular area, the headquarters will first of all consider the factors of political and physical geography and the presence of persons within the area who have access to the desired information. Obviously, contiguous and border areas around the great periphery of the Communist world serve as windows, though darkly shaded ones, on that world. The presence of sizable delegations from the Sino-Soviet bloc in many countries not necessarily contiguous to it offers quite another kind of opportunity for information on the bloc. Also, citizens of peripheral countries may not have the difficulties an American would have in traveling to denied areas and enjoying more freedom of movement and less close scrutiny while there. All these are factors in the problem of "access" and therefore play a role in the framing of guidance.

Hypothetically speaking, if our government wanted information on a recent industrial or technical development in Red China, where the U.S. has no diplomatic mission and no unofficial representation either, the intelligence service could assign the collection task to those free areas close to China which receive Chinese refugees from time to time, or to a free area halfway around the world from China where the latter had a diplomatic mission, or to still another free area which had commercial relations with China and whose nationals could travel there. It would not assign the task to an area where none of these conditions existed, nor would it indiscriminately flash out its requirement worldwide, setting up a scramble of intelligence officers to go after the same information by whatever means they could devise.

When Khrushchev made his secret speech denouncing Stalin to the Twentieth Party Congress in 1956, it was clear from various press and other references to the speech that a text must be available somewhere. The speech was too long and too detailed to have been made extemporaneously even by Khrushchev, who is noted for lengthy extemporary remarks. An intelligence "document hunt" was instituted, as the speech, never published in the U.S.S.R., was of great importance for the Free World. Eventually the text was found—but many miles from Moscow, where it had been delivered. It was necessary in this case for headquarters to alert many kinds of sources and to make sure all clues were followed up. I have always viewed this as one of the major coups of my tour of duty in intelligence. Since the text was published in full by the State

Department, it also was one of the few exploits which could be disclosed as long as sources and methods of acquisition were kept secret.

Usually the means of getting the information once a task has been assigned is left to the ingenuity of the intelligence officer in the field. My source in the German Foreign Office already mentioned brought out or secretly smuggled to me in Switzerland during 1943–45, choice selections of the most secret German diplomatic and military messages, over two thousand in all. For various technical reasons, he could send only a fraction of the total available to him, and he had to pick and choose on his own initiative.

As the war in Europe was drawing to a close, the possibility of a protracted conflict with Japan still loomed ahead. I then received from headquarters a request that our source concentrate on sending me more reports from German missions in the Far East, particularly in Tokyo and Shanghai. Even though I agreed with headquarters that this window on the Far East should be opened wider, it was no easy task to carry out the instruction speedily.

My source was in Berlin and I was in Switzerland. He was able to travel out only rarely, I might not see him for weeks, and the matter was too urgent to let go until our next meeting. Normally we never communicated with each other across the Swiss-German border because it was too dangerous, but we did have an emergency arrangement based on a fictitious girl friend of the source who was supposedly living in Switzerland. Since postcards seem more innocent to the censor than sealed letters, the "girl friend" sent to the source's home address in Berlin a beautiful postal card of the Jungfrau. "She" wrote on it that a friend of hers in Zurich had a shop which formerly sold Japanese toys but had run out of them and couldn't import them because of wartime restrictions; in view of the close relations between Germany and Japan, couldn't he help her out by suggesting where in Germany she could buy Japanese toys for her shop? My source got the point immediately since he knew all messages from the Swiss "girl friend" were from me. The next batch of cables to the German Foreign Office which he sent me were largely from German officials in the Far East and told the plight of the Japanese Navy and Air Force.

Sometimes for diplomatic or other reasons an intelligence headquarters gives out negative guidance, i.e., instructions what not to do. An enterprising intelligence officer may run into some splendid opportunities and learn to his disappointment after corresponding with his headquarters that there are good reasons for passing them up. He may or may not be told what these good reasons are.

General Marshall, in the letter to Governor Dewey mentioned earlier, emphasized the sensitivity of operations involving enemy codes and ciphers by telling him of an uncoordinated attempt by American intelligence to get a German code in Portugal. The operation misfired and so alerted the Germans that they changed a code we were already reading, and this valuable source was lost.

I had no knowledge of this incident at the time when I received a radio message from headquarters at my wartime post in Switzerland not to get *any* foreign codes without prior instructions. Shortly after this, in late 1944, one of my most trusted German agents told me that he could get me detailed information about certain Nazi codes and ciphers. This put me in quite a quandary. Though I had confidence in him, I did not wish him to deduce that we were breaking the German codes. If I showed no interest, this would have been an indication that such was the case. No intelligence officer would otherwise reject such an offer. I told my friend I wanted a bit of time to think over how best this could be worked out. The next day I told him that as all my traffic to Washington had to go by radio—Switzerland was then surrounded by Nazi and Fascist forces—it would be too insecure for me to communicate what he might give me. I said I preferred to wait till France was liberated—the Normandy invasion had already taken place—so I could send out his code information by diplomatic pouch. He readily accepted this somewhat specious answer.

The best planning and the best guidance cannot, of course, foresee everything. No intelligence service and no intelligence officer rules out the possibility of the random and unexpected and often inexplicable windfall. Sometimes a man who has something on his mind feels safer talking to a Western intelligence officer ten thousand miles from home and so waits for the opportunity of a trip abroad to seek one out. A Soviet scientist or technician visiting Southeast Asia, for example, might talk in a more relaxed manner than if he were behind the Curtain or even if he were visiting in New York. The Kremlin's instruction to a Soviet official in Egypt, if it came to our attention, might throw some light on Soviet policy toward Berlin.

In 1958 an Arab student from Iraq who had been taking some advanced studies in Arizona received a letter from Baghdad which caused him to leave immediately for home. As he departed, he hinted to an

American friend of his that the reason for his sudden leave-taking was that important political events were impending in his home country. A few weeks later came the Iraq *coup d'état* which astounded the Western world and left some intelligence officers with red faces. This bit of information about the student's hasty departure, and the reason for it, thanks to some good work of field collection in Arizona did in fact reach headquarters in Washington quite promptly. Unfortunately, there it was viewed at the desk level, and quite naturally, as only one straw in a wind which seemed to be blowing in a different direction.

This story also illustrates how important it is for the field officer, without any directives or headquarters administration, to send in bits and pieces of intelligence. If, for example, in the Iraq case, headquarters had received three or four messages that persons at "outs" with the Iraq government were converging toward Baghdad, a quiet alert should have been sounded.

Some years ago, when the Moscow meetings of the Central Committee of the Communist part were often held in great secrecy, they could sometimes be predicted by noting the movements of the many committee members serving in diplomatic or other posts or traveling abroad. If they quietly converged on Moscow, as they did just before the ouster of Khrushchev, something was likely to be about to happen. Here the travel pattern of Soviet officials was a type of information which field officers were alerted to follow.

Headquarters guidance is necessary but it is no substitute for such field initiative as was taken in Arizona.

7

The Main Opponent —
The Communist Intelligence Services

Most totalitarian countries have, in the course of time, developed not just one but two intelligence services with quite distinct functions, even though the work of these services may occasionally overlap. One of these organizations is a military intelligence service run by the general staff of the armed forces and responsible for collecting military and technical information abroad. In the U.S.S.R. this military organization is called the GRU (Main Intelligence Directorate). GRU officers working out of the Soviet Embassy in Ottawa operated the atomic spy networks in Canada during World War II. The other service, which more typically represents an exclusive development of a totalitarian state, is the "security" service. Generally such a service has its origin in a secret police force devoted to internal affairs such as the repression of dissidents and the protection of the regime. Gradually this organization expands outward, thrusting into neighboring areas for "protective" reasons, and finally spreads out over the globe as a full-fledged foreign intelligence service and much more.

Since this security service is primarily the creation of the clique or party in power, it will always be more trusted by political leaders than is the military intelligence service, and it will usually seek to dominate and control the military service, if not to absorb it. In Nazi Germany the "Reich Security Office," under Himmler, during 1944 completely took over its military counterpart, the *Abwehr*. In 1947, the security and military services in Soviet Russia were combined, with the former dominant, but the merger lasted only a year. In 1958, however, Khrushchev placed one of his most trusted security chiefs, General Ivan Serov, in charge of the GRU, apparently in order to keep an eye on it. It was Serov, one of the most brutal men in Soviet intelligence history, whom Khrushchev called upon to direct the suppression of the Hungarian Revolution and the Soviet "reconquest" of Hungary in November of 1956. There are, incidentally, indications that things have not gone too well for Serov, that he was caught up in one of the dramatic housecleanings that so often sweep through the Soviet security services.

Whether or not the security service of a totalitarian state succeeds in gaining control of the military service, it inevitably becomes the more powerful organization. Furthermore, its mandate, both internal and external, far exceeds that of the intelligence services of free societies. Today the Soviet State Security Service (KGB) is the eyes and ears of the Soviet state abroad as well as at home. It is a multipurpose, clandestine arm of power that can in the last analysis carry out almost any act that the Soviet leadership assigns to it. It is more than a secret police organization, more than an intelligence and counterintelligence organization. It is an instrument for subversion, manipulation and violence, for secret intervention in the affairs of other countries. It is an aggressive arm of Soviet ambitions in the Cold War. If the Soviets send astronauts to the moon, I expect that a KGB officer will accompany them.

No sooner had the Bolsheviks seized power in Russia than they established their own secret police. The Cheka was set up under Feliks Dzerzhinski in December, 1917, as a security force with executive powers. The name stood for "Extraordinary Commission against Counter-Revolution and Sabotage." The Cheka was a militant, terroristic police force that ruthlessly liquidated civilians on the basis of denunciations and suspicion of bourgeois origins. It followed the Red armies in their conflicts with the White Russian forces, and operated as a kind of counterespionage organization in areas where sovietization had not yet been accomplished. In 1921 it established a foreign arm, because by that time White Russian soldiers and civilian opponents of the Bolsheviks who could manage to do so had fled to Western Europe and the Middle and Far East and were seeking to strike back against the Bolsheviks from abroad.

Almost at once this foreign arm of Soviet security had a much bigger job than ever confronted the Czar's Okhrana. It had not only to penetrate and neutralize the Russian exile organizations that were conspiring against the Soviets, but also to watch, and wherever possible to influence, the Western powers hostile to the Bolsheviks. It thus became a political intelligence service with a militant mission. In order to achieve its aims, it engaged in violence and brutality, in kidnaping and murder, both at home and abroad. This activity was directed not only against the "enemies of the state," but against fellow Bolsheviks who were considered untrustworthy or burdensome. In Paris, in 1926, General Petlura, the exiled leader of the Ukrainian nationalists, was murdered; some say it was by the security service, others claim it was personal vengeance. In 1930, again in Paris, the service kidnaped General Kutepov, the leader of the White Russian war veterans; in 1937 the same fate befell his successor, General Miller. For over a decade Leon Trotski, who had gone into exile in 1929, was the prime assassination target of Stalin. On August 21, 1940, the old revolutionist died in Mexico City after being slashed with an Alpine climber's ice ax by an agent of Soviet security. The list of its own officers and agents abroad whom it murdered during this same period, many of whom had tried to break away or were simply not trusted by Stalin, is far longer.

Lest anyone think that violent acts against exiles who opposed or broke with the Bolsheviks in the early days were merely manifestations of the rough-and-tumble era of early Soviet history or of Stalin's personal vengefulness, it should be pointed out that in the subsequent era of so-called "socialist legality," which was proclaimed by Khrushchev in 1956, a later generation of exiled leaders was decimated. The only difference between the earlier and later crops of political murders lay in the subtlety and efficacy of the murder weapons. The mysterious deaths in Munich, in 1957 and 1959, of Lev Rebet and Stephen Bandera, leaders of the Ukrainian émigrés, were managed with a cyanide spray that killed almost instantaneously. This method was so effective that in Rebet's case it was long thought that he had died of a heart attack. The truth became known only when the KGB agent Bogdan Stashinski gave himself up to the German police in 1961 and acknowledged that he had perpetrated both killings.

For the first murder, Stashinski reports he was given a fine banquet by his superiors in the KGB; for the second he received from them the Order of the Red Banner.

Since the earliest days of the Soviets, secret assassination has been an official state function assigned to the apparatus of the security service. A special "Executive Action" section within the latter has the responsibility for planning such assassinations, choosing and training the assassin, and seeing to it that the job is carried out in such a way that the Soviet government cannot be traced as the perpetrator. That this section is still today a most important component of Soviet intelligence is borne out by the fact that General Korovin[1] has been serving as its chief. While counselor of the Soviet Embassy in London from 1953 until early 1961, he was in charge of two key Soviet spies in Britain, George Blake and William John Vassall. After the apprehension of the latter, the ground got too hot for the General and he was recalled and reassigned to the "Executive Action" branch of the KGB.

EVOLUTION OF SOVIET SECURITY SERVICES

In 1922 the Cheka became the GPU (State Political Administration), which in 1934 became part of the NKVD (People's Commissariat for Internal Affairs). This consolidation finally brought together under one ministry all civilian security and intelligence bodies—secret, overt, domestic and foreign. As the foreign arm of Soviet security was expanding into a world-wide espionage and political action organization, the domestic arm grew into a monster. It is said that under Stalin one out of every five Soviet citizens was reporting to it. In addition, it exercised control over the entire border militia, had an internal militia of its own, ran all the prisons and labor and concentration camps, and had become the watchdog over the government and over the Communist party itself. Its most frightening power as an internal secret police lay in its authority to arrest, condemn and liquidate at the behest of the dictator, his henchmen or even on its own cognizance, without any recourse to legal judgment or control by any other organ of government.

During the war years and afterward the colossus of the NKVD was split up, reconsolidated, split up again, reconsolidated again and finally split up once more into two separate organizations. The MGB, now KGB, was made responsible for external espionage and high-level internal security; the other organization retained all policing functions not directly concerned with state security at the higher levels and was called the MVD (Ministry of Internal Affairs).

[1]This was the alias used by the General while in London. His real name is Nikolay B. Rodin.

Obviously, any clandestine arm that can so permeate and control public life, even in the upper echelons of power, must be kept under the absolute control of the dictator. Thus it must occasionally be purged and weakened to keep it from swallowing up everything, the dictator included. The history of Soviet state security, under its various names, exhibits many cycles of growing strength and subsequent purge, of consolidation and of splintering, of rashes of political murders carried out by it and sometimes against it.

After any period during which a leader had exploited it to keep himself in power, it had to be cut down to size, both because it knew too much and because it might become too strong for his own safety. After the demise of a dictator, the same had to be done for the safety of his successor.

Stalin used the GPU to enforce collectivization and liquidate the kulaks during the early thirties, and the NKVD during the mid-thirties to wipe out all the people he did not trust or like in the party, the army and the government. Then in 1937 he purged the instrument of liquidation itself. Its chiefs and leading officers knew too much about his crimes, and their power was second only to his. In 1953, after the death of Stalin, the security service was again strong enough to become a dominant force in the struggle for power, and the so-called "collective leadership" felt they would not be safe until they had liquidated its leader, Lavrenti Beria, and cleaned out his henchmen.

In Khrushchev's now famous address to the Twentieth Congress of the Communist party in 1956, in which he exposed the crimes of Stalin, the main emphasis was on those crimes Stalin had committed through the NKVD. This speech not only served to open Khrushchev's attack on Stalinism and the Stalinists still in the regime, but was also intended to justify new purges of existing state security organs, which he had to bring under his control in order to strengthen his own position as dictator. Anxious to give both the Soviet public and the outside world the impression that the new era of "socialist legality" was dawning, Khrushchev subsequently took various steps to wipe out the image of the security service as a repressive executive body. One of these was the announcement on September 3, 1962, that the Ministry of Internal Affairs (MVD) was now to be called the Ministry of Public Law and Order. Just what this new ministry would do he did not clarify, although he did promise that no more trials would be held in which Soviet citizens were condemned in secret.

Yet internal control systems still exist, even though in new forms. For example, under the terms of a decree published on November 28, 1962, an elaborate control system has been established which, to quote the *New York Times* (November 29, 1962), "would make every worker in every job a watchman over the implementation of party and government directives." In commenting on the decree *Pravda* made reference to earlier poor controls over "faking, pilfering, bribing and bureaucracy," and asserted that the new system would be a "sharp weapon" against them, as well as against "red tape and misuse of authority" and "squanderers of the national wealth." The new watchdog agency is called the Committee of Party and State Control.

With so many informers operating against such broad categories of crimes and misdemeanors, it should be possible to put almost anyone in jail at any time. And indeed the press has been full of reports recently that courts in the Soviet Union have been handing down death or long prison sentences for many offenses that in the United States would be only minor crimes or misdemeanors.

On February 5, 1963, we learned for example that the director and manager of the Sverdlovsk railway station restaurant had been condemned to death by the court in Sverdlovsk for inventing and using a machine for frying meat and pies which required two or three grams less fat than regulations called for. The two men pocketed the difference and swindled the government out of four hundred rubles monthly. There is something alarmingly out of joint in a country that today will levy the death penalty for such crimes and calls for the collaboration of the ordinary citizen with the secret police in order to discover them. Aleksandr N. Shelepin, who was designated by the Central Committee of the Communist party of the Soviet Union to be the head of this new control agency, once served as head of the KGB, having succeeded General Ivan Serov in 1958.

But all these shake-ups, purges and organizational changes seem to have had remarkably little effect on the aims, methods and capabilities of that part of the Soviet security service which interests us most—its foreign arm. Throughout its forty-five years this world-wide clandestine apparatus has accumulated an enormous fund of knowledge and experience; its techniques have been amply tested for their suitability in furthering Soviet aims in various parts of the world, and its exhaustive files of intelligence information have been kept intact through all the political power struggles. It has in its ranks intelligence officers (those who survived the purges) of twenty to thirty years' experience. It has on its rosters disciplined, experienced agents and informants spread throughout the world, many of whom have been active since the 1930s. And it has a tradition that goes all the way back to czarist days.

On December 20, 1962, an article appeared in *Pravda* under the name of the Chief of Soviet State Security (KGB), M. Semichastny, which opened with the words, "Forty-five years ago today, at the initiative of Vladimir Illitch Lenin . . ." and went on to describe the founding of the first Soviet security body, the Cheka, in 1917, and to summarize the ups and downs of forty-five years of Soviet police and intelligence history. While the purpose of the article was no doubt to improve the public image of this justly feared and hated institution, its importance to the foreign observer lay in the tacit admission that despite changes of name and of leadership the Soviets really view this organization as having a definite and unbroken continuity since the day of its founding.

In their attempts to evade detection and capture by the Okhrana, the Russian revolutionaries of the late nineteenth and early twentieth centuries developed the conspiratorial techniques that later stood the Soviets in such good stead. The complicated and devious tricks of concealing and passing messages, of falsifying documents, of using harmless intermediaries between suspect parties so as not to expose one to the other or allow both to be seen together—these were all survival techniques developed after bitter encounters and many losses at the hands of the czar's police. When the Soviets later founded their own intelligence service, these were the tricks they taught their agents to evade the police of other countries. Even the very words which the Bolsheviks used in the illegal days before 1917 as a kind of private slang among terrorists—such as *dubok* (little oak tree) for a dead-letter drop—became in time the terms in official use within the Soviet intelligence service.

It is always a matter of surmise among Western observers whether the internal power struggles which are usually rife within the hierarchy of the Soviet Union will affect the position and powers of the KGB as the most privileged body in the Soviet state. I do not mean solely that its top people may be removed, or even executed, as were the former chiefs, Yezhov, Yagoda and Beria, in their day, but rather that its entire ranks might be purged and its standing vis-à-vis other elements of the state sharply reduced. The chief contender for power is the Army, which time and again in Soviet history has been downgraded by the dictator in favor of the state security organization, since the latter was his personal instrument and he could use it to keep an eye on the army.

THE INTELLIGENCE SERVICES OF THE EUROPEAN SATELLITES AND RED CHINA

Soviet State Security founded, organized, trained and today still supervises the intelligence and security services of the European satellites of Soviet Russia. They are in a sense little "KGBs" and sometimes like to call themselves that within their own ranks. They are entirely the creatures of the Soviets and mirror in their structure and their techniques the results of the long-range experience of their Soviet big brothers. Their main objectives are dictated by the Soviets, although they are allowed certain limited initiatives in matters relating to their own "internal" security. The Poles and Czechs, for example, will run operations whose intent is to locate Western espionage directed against their national areas. If in the course of such operations they turn up an especially good agent who offers, let us say, a prime opportunity for penetration of a Western government, the Soviets will very likely take over the agent and run him themselves, and the satellite intelligence service must grin and bear it.

This was the case with Harry Houghton, who was first recruited by the Polish intelligence service when he was stationed at the British embassy in Warsaw. When he was transferred back to London and put to work in the Admiralty, the Soviets saw opportunities which were far too important for the Poles to handle. They took over the case and the Polish intelligence service never heard about Houghton again until his name appeared in the papers after his arrest.

From the beginning the Soviets maintained an efficient stranglehold over these services by appointing to the top jobs in them people who had been old-line Soviet agents and had been trained in Moscow, many of them in pre-World War II days. The hard core of the present Polish intelligence service, for example, is made up of Polish Communists who had fled to Russia in 1939 and who returned to Poland in 1944 with Polish military units accompanying the Red Army. They had spent most of the war years in Moscow being trained by the Soviets for their future jobs in a projected but as yet nonexistent Polish intelligence service. Younger personnel are carefully screened by the Soviets before being accepted for employment in any of the satellite services.

Even today the Soviets manage and direct the satellite services, not at long range but in person. They do this through a so-called advisory system. A Soviet "adviser" is installed in almost every significant department of the satellite intelligence services, be it in Prague, Warsaw, Bucharest or any other satellite capital. This adviser is supposed to be shown all significant material concerning the work being done, and must give consent to all important operational undertakings. He is to all intents and purposes a supervisor, and his word is final.

As a sidelight on Soviet relations to the satellites, it is interesting to note that the Soviets do not rely wholly on these advisers to control the satellite intelligence services. This is not because the latter are incompetent, but because the satellite services are evidently not trusted by their Soviet master. In order to prevent these services from getting away with anything, the Soviets go to the trouble of secretly recruiting intelligence officers of the satellite services who can supply them with information on plans, personnel, conflicts in the local management, disaffection and the like, which might not have come to the attention of the adviser.

While the Soviets cannot really trust their satellites, they will use them to draw chestnuts out of the fire where it is advantageous to do so. The Soviets were quick to recognize, for example, that the very great numbers of persons of Polish, Czech and Hungarian extraction living in Western Europe and in Canada and the United States theoretically represented a potential pool of agents to which the respective satellite services might find access with much greater ease than the Soviets could, on the basis of common ethnic background, family and other sentimental ties to the old country, etc. Thus, we find that the attempts to recruit people of Central European and Balkan extraction both here and abroad for Communist espionage have largely been carried out by personnel of the satellite intelligence services. That the latter have been rebuffed in most cases is a tribute to the loyalties of the first- or second-generation citizens of the U.S. and the other NATO countries.

Red China, not being a satrapy of the Soviet Union as are the smaller nations of Eastern Europe, has its own independent intelligence and security system which is in no way subservient to the KGB. In intelligence as in technical and scientific fields, the Soviets for a long period had advisers stationed in China, but these were really advisers and not the kind of supervisors I described above. They have long since departed, and it is unlikely today, in view of the Sino-Soviet rift, that there is more than the most nominal collaboration and coordination between the Red Chinese and the Soviet intelligence services. Indeed, we can safely assume that each of these countries is using its intelligence service to keep its eye on the other.

We have not yet begun to consider Red Chinese espionage as a serious menace to our own security in the U.S., though in the years to come it may well become a formidable instrument for spying and subversion in the West, as it already has throughout Asia and the Pacific. The Chinese are, of course, at the same disadvantage in operating against us as we are in attempting to operate against them. Physical and cultural differences make it quite difficult to camouflage the true ethnic status and national origin of intelligence officers or agents on either side.

A Ukrainian was able with sufficient training and with the proper documents to pass himself off in England as a Canadian of Anglo-Saxon origin named Gordon Lonsdale. For a Chinese, this would, of course, be impossible. In areas where there are large numbers of resident Chinese, as in Hawaii, Malaya, etc., the Chinese can take advantage of ethnic ties. The first real inroads into Occidental areas are now being made by the Chinese in South America, where the more fanatical element of the local Communist contingents welcomes them. Should the Chinese succeed in such areas in recruiting Westerners of Hispanic origin as long-term agents, it will begin to be possible for them to infiltrate the U.S. and European countries with such agents, who would be no more recognizable as Chinese agents than Lonsdale was as a Soviet agent.

There is reason to expect an ever greater effort on the part of Red Chinese intelligence against U.S. and other Western targets. China is anxious to develop its nuclear power, but the withdrawal of Soviet technical advisers in 1959 undoubtedly slowed down its program in this field. The course of the Red Chinese will very likely be the same the Soviets chose during and after World War II, when they succeeded in stealing atomic secrets from us through spies like Fuchs and Pontecorvo and penetrated American and other Western scientific installations. J. Edgar Hoover, Director of the FBI, warned American industrialists in early 1964 that the Red Chinese were seeking to gain information on American technical installations through the use of Chinese-Americans long established in this country and also by exploiting the contacts dating from college days which Chinese scientists trained in the United States formerly had with American scientists. Should the Red Chinese be admitted to the U.N. or establish diplomatic installations on our soil, they would then have firmer bases from which to organize and direct their technical espionage undertakings.

In the Western European countries that have recognized Red China diplomatically, among which France can now be numbered, the Chinese have staffed their embassies with a quantity of personnel far in excess of the normal and necessary contingents and with unusually frequent turnovers of such personnel. This has been the case, for example, in Bern, Switzerland, where the Chinese have well over a hundred employees stationed, obviously many more than are needed for the normal course of their diplomacy with Switzerland. What percentage of these are engaged in intelligence work is not easy to determine. It is clear, however, that many of them are sent abroad solely to learn Western ways and to become acquainted with the workings of Western societies and enterprises, doubtless as part of their training for future intelligence work.

THE SOVIET INTELLIGENCE OFFICER

From my own experience I have the impression that the Soviet intelligence officer represents the species *homo Sovieticus* in its unalloyed and most successful form. This strikes me as much the most important thing about him, more important than his characteristics as a practitioner of the intelligence craft itself. It is as if the Soviet intelligence officer were a kind of final and extreme product of the Soviet system, an example of the Soviet mentality pitched to the nth degree.

He is blindly and unquestioningly dedicated to the cause, at least at the outset. He has been fully indoctrinated in the political and philosophical beliefs of Communism and in the basic motivation which proceeds from these beliefs, which is that the ends alone count and any means which achieve them are justified. Since the ingrained Soviet approach to the problems of life and politics is conspiratorial, it is no surprise that this approach finds its ultimate fulfillment in intelligence work. When such a man does finally see the light, as has happened, his disillusionment is overwhelming.

The Soviet intelligence officer is throughout his career subject to a rigid discipline; as one intelligence officer put it who had experienced this discipline himself, he "has graduated from an iron school." On the one hand, he belongs to an elite and has privilege and power of a very special kind. He may be functioning as the embassy chauffeur, but he may have a higher secret rank than the ambassador and more power where the power really counts. On the other hand, neither rank nor seniority nor past achievement will protect him if he makes a mistake. When a Soviet intelligence officer is caught out or his agents are caught through an oversight on his part, he can expect demotion, dismissal, even prison. In Stalin's day he would have been shot.

I can think of no better illustration of the merciless attitude of the Soviet intelligence officer himself than the story told of one of Stalin's intelligence chiefs, General V. S. Abakumov. During the war, Abakumov's sister was picked up somewhere in Russia on a minor black-marketing charge—"speculation," as the Soviets call it. In view of her close connection to this powerful officer in the secret hierarchy, the police officials sent a message to Abakumov asking how he would like the case handled. They fully expected he would request the charges be dropped. Instead, he is reliably reported to have written on the memorandum sent him: "Why do you ask me? Don't you know your duty? Speculation during wartime is treason. Shoot her." An interesting sidelight on Abakumov is that he, like his boss, Beria, ran what one writer has described as "a string of private brothels."

Abakumov met the fate of many Soviet intelligence officers after the death of Stalin and the liquidation of Beria. At that time he was in charge of the internal section of Soviet security, which kept the files on members of the government and of the party. Abakumov was secretly executed and his entire section was decimated under the Malenkov regime. They knew too much. Despite certain relaxations in the public life of present-day Russia, the "terror" still holds sway within Soviet intelligence itself because this arm of Soviet power, second to none in peacetime, cannot relax, cannot be allowed any weakness.

In Soviet Russia, where the foreign intelligence service and the internal secret police at the higher levels are only separate arms of the KGB, most officers rotate between the two different types of duty. They customarily are assigned early in their careers to some provincial secret police office, usually in an area of their country in which they are not native. Here their duties primarily call for the running of informants among the local populace. Besides carrying out a function which the Soviet state deems necessary for its own internal security, men working at such posts also receive a basic on-the-job training in the fundamentals of espionage and counterespionage and at a level where occasional errors are not especially damaging.

Less gifted officers may remain at such posts for the greater part of their careers. The better men will eventually be assigned to intelligence headquarters. When they have sufficient experience and are thought to have been adequately tested for trustworthiness from the Communist point of view, they may finally be sent to a foreign post.

Peter Deriabin, who came over to us in Vienna in 1954, relates in his book that he began his KGB career with an assignment to the section responsible for guarding the lives of the Soviet bigwigs.[2] He spent five years in this section and finally succeeded in getting himself assigned to a branch of the Foreign Intelligence Department responsible for operations in Austria. This, as would be the case in most intelligence services, gradually opened the way for his own transfer to a foreign post, logically enough, in Vienna. But he had served in the KGB over six years before he was entrusted with a foreign assignment.

The Soviets prefer to send men abroad who have had counter-intelligence experience within Soviet Russia, and for a noteworthy reason. Having sat for years in posts where their primary responsibility was apprehending opponents of the regime, penetrating dissident circles and tracking an occasional miscreant suspected of cooperating with the "imperialists," they are well aware of the workings of the secret police mentality. When the tables are turned and they find themselves in foreign countries running their own spy networks, they are likely to anticipate and often to outwit local police organs for whom they now represent the potential victim.

After returning from a tour of duty abroad in which they did not especially distinguish themselves, they may be assigned again to provincial police duties. The Soviets thus have a built-in solution for disposing of superannuated or ineffective intelligence officers. If, on the other hand, they did well abroad, they may begin to go up the administrative ladder in the foreign intelligence department, which is the most preferred and privileged branch of the service.

The Soviet citizen does not usually apply for a job in the intelligence service. He is spotted and chosen. Bright upcoming young men in various positions, be it in foreign affairs, economics or the sciences, are proposed by their superiors in the party for work in intelligence. To pass muster they must either be party members themselves, candidates for

[2] Peter Deriabin and Frank Gibney, *The Secret World* (New York: Doubleday & Co., Inc., 1959).

party membership or members of the youth organization, *Komsomol*, which is a kind of junior Communist party. They must come from an impeccable political background according to Communist standards, which means that there can be no "bourgeois taint" or any record of deviation or dissent in their immediate family or forebears.

An ambitious young man who is able to make his career in one of the branches of Soviet intelligence is fortunate by Soviet standards. His selection for this duty opens to him the doors of the "New Class," the elite, the nobility of the new Soviet state. Soviet intelligence officers are ranked, as are the military, and have the same titles, although they only use these titles within the service at home. Rudolf Abel, who so successfully acted the part of a second-rate photographer in Brooklyn, was a colonel in Soviet intelligence. The heads of large departments usually rank as major- or lieutenant-generals. But service with Khrushchev's security and intelligence often surpasses the prestige of service with the military. Soviet intelligence officers receive material rewards much above those given the similar ranks of government bureaucracy in other departments. They have opportunities for travel open to few Soviet citizens. Further, a career of this kind may open the road to high political office and important rank in the Communist party.

This is the breed of men who handled such cases as Chambers and Klaus Fuchs, the Rosenbergs, Burgess and Maclean, George Blake, Houghton and Vassall. They have had some brilliant successes. What are their weaknesses and shortcomings?

The Soviet Security Service suffers from the same fundamental weakness as does Soviet bureaucracy and Communist society generally—indifference to the individual and his feelings, resulting in frequent lack of recognition, improper assignments, frustrated ambition, unfair punishment, all of which breed, in a Soviet Russian as in any man, loss of initiative, passivity, disgruntlement and dissidence. Service in the Soviet bureaucracy does not exactly foster independent thought and the qualities of leadership. The average Soviet official, in the intelligence service as elsewhere, is not inclined to assume responsibility or risk his career. There is an ingrained tendency to perform tasks "by the book," to conform, to try to pass the bureaucratic buck if things go wrong.

Most important of all, every time the Soviets send an intelligence officer abroad they risk his exposure to the very systems he is dedicated to destroy. If for any reason he has become disillusioned or dissatisfied, his contact with the Western world often works as the catalyst which starts the process of disaffection. A steady and growing number of Soviet intelligence officers have been coming over to our side, proving that Soviet intelligence is by no means as monolithic and invulnerable as it wishes the world to believe.

SOME SOVIET TECHNIQUES—LEGALS AND ILLEGALS

I have already referred to "illegals" in an earlier chapter as a kind of "made-over" man. In Soviet practice not only agents but the staff intelligence officer himself may go abroad as an illegal. In the 1920s, when the Soviets ran their intelligence operations out of their diplomatic establishments abroad, these operations, which at that time were by no means particularly sophisticated, frequently fell afoul of the local police with the result that the espionage center was traced down to the local Soviet embassy, forcing the recall of the intelligence personnel stationed there and often harming Soviet relations with important countries, such as France and England with whom the Soviets for economic and other reasons wished to stay on outwardly good terms. It was at this time, in an attempt to keep espionage and diplomacy ostensibly separate, with advantages for both, that the Soviets hit upon the idea of developing a duplicate espionage apparatus in each country. Within the embassy there would still be intelligence officers but they would restrict themselves, except in emergencies, to "clean" operations, of which I have more to say below. This unit the Soviets call the "legal *residentura*." Outside the embassy and buried away under the guise of some harmless occupation, perhaps in a bookstore or a photography shop, was quite another center devoted to the "dirty" operations. This was headquarters of the "illegal *residentura*," composed mainly of officers who over a period of years had carefully been turned into personages whom it would be almost impossible to identify as Soviet nationals, much less as intelligence personnel. The illegal, unless apprehended with the agent or betrayed by him, can disappear into the woodwork if something goes wrong. There will be no trail leading to a Soviet diplomatic installation to embarrass or discredit it. A principle governing this double setup was that neither center would have anything at all to do with the other except in emergencies. Each had its own separate communications with Moscow and only took its orders from there. The legal *residentura* used diplomatic channels of communication. The illegals had their own radio operators, a most dangerous and difficult arrangement. Most of the great Soviet wartime intelligence nets, as we shall see, came to grief because of their secret radio communications.

A man chosen for illegal work in any of its aspects will be sent to live abroad for as many years as it takes him to perfect his knowledge of the language and way of life of another country. He may even acquire citizenship in the adopted country. But during this whole period he has absolutely no intelligence mission. He does nothing that would arouse suspicion. When he has become sufficiently acclimatized, he returns to the Soviet Union, where he is trained and documented for his intelligence mission, and eventually dispatched to the target country, which may be the same one he has learned to live in or a different one. It matters little, for the main thing is that he is unrecognizable as a Soviet or Eastern European. He is a German or a Scandinavian or a South American. His papers show it, and so do his speech and his manners.

Sometimes, to provide their illegals with documents, the Soviets make use of the papers of a family which has been wiped out. For example, after the liberation of the Baltic states in World War I, many Americans of Lithuanian extraction returned to their native habitat with their children. Two decades later, when the Baltic states were overrun by the Soviets, many of these people were caught in the liquidation of anti-Communists which followed. Their papers, including the birth certificates of their American-born children, fell into the hands of the Soviet police. Later the KGB found these extremely useful for documenting their agents with bona fide American passports.

In most Western countries lax procedures in the issuance of duplicate birth certificates, records of marriage, death, etc., make it relatively easy for hostile intelligence services to procure valid documents for "papering" their agents. This situation has been frequently used by the Soviets, and any measures taken to correct it would be of distinct service to Western security.

Because they have almost perfect camouflage and are consequently immensely difficult to locate, "illegals" constitute the gravest security hazards to countries against which they are working. There is every evidence that the Soviets have been turning out such "illegals" at an accelerated rate since the end of World War II. Generally, they are used in a supervisory capacity, for directing espionage networks, rather than for penetration jobs that increase the danger of discovery.

However, despite the lengths to which the Soviets go to create illegals, a number of them of major stature have been uncovered and apprehended by Western intelligence in recent years. In 1957 Colonel Rudolf Abel, alias Emil R. Goldfus, was caught in the United States. He was tried and sentenced but was exchanged in 1962, after serving five years in prison, for the downed U-2 pilot, Francis Gary Powers. In early 1961 the British caught Conon Molody, alias Gordon Londsdale, in London and with him four other Soviet agents in what became known as the Naval Secrets Case. Lonsdale spoke perfect English and passed for a small-time businessman dealing in jukeboxes. His Canadian identity had been built up over many years, but the Soviets used him not in Canada, where he would have been exposed to accidental encounters with people from his "home town," but in England, where, as a Canadian, he would be quite acceptable and would be unlikely to become the subject of much curiosity about the details of his background.

When an intelligence service goes to all the trouble to retool and remake a man so that he can succeed in losing himself in the crowd in another country, it naturally does so in the expectation that the man will stay put and remain active and useful for a long period of time. There is no rotation here of the sort that is common among officials of most diplomatic and intelligence services. Also, for obvious reasons, if the "illegal" has a family, the family does not accompany him. The wives and children cannot also be "made over." He goes alone, and even his communications to his wife and children must necessarily be limited and must pass through secret channels. The only glimpse of Colonel Abel as a human being, indeed the only glimpse of the man as anything but a tight-lipped automaton, was afforded by some letters found in his possession which were written by his wife and daughter. Abel had been at his post nine years when he was caught. There is no reason to believe that he would not have continued in it for many years if one of his fellow workers, also an illegal, had not turned himself over to the U.S.A.

There are times, of course, when the "cover" of the embassy or trade mission lends advantages to the "legal" center not available to the illegal. Under the guise of "business" or "social" relations an officer in an embassy may be able to make certain connections in circles to which he has access which would be denied to the illegal.

If the Soviets, for example, are anxious to find an agent in a Western country who can report to them on a sensitive industry, the Soviet Trade Mission will advertise that it is interested in purchasing certain nonstrategic items manufactured by that industry or one closely allied to it. Manufacturers or middlemen will be attracted by the ad and will visit the Soviet mission to talk over possible business. They will be requested to fill out forms that call for personal and business data, references, financial statements, etc.

All this material is reviewed by the intelligence officer stationed at the mission. If any candidates seem promising because of their innocence, their political or perhaps apolitical attitudes, their need for money or susceptibility to blackmail, the Soviets can cultivate them further by pretending that the business deal is slowly brewing. The hand of espionage has not yet been shown. Nothing ostensibly has yet been done against the law.

Similarly, if Soviet intelligence officers stationed at an embassy and belonging to the legal *residentura* meet interesting or influential persons from the local environment in the course of the dinners, parties or other social events (which the Soviets now give in order to create a certain sophisticated and "friendly" impression in contrast to their behavior in earlier decades), they may very likely develop these "friendships" and even risk a recruitment at a later date. However, some of their recent attempts of this sort, particularly through their UN personnel, have been so crude and bare-faced as to make one wonder whether the Soviets are not using the UN for the schooling of their intelligence officers. It is also apparent from some recent cases that the Soviets have not been able to establish "illegals" in certain countries and therefore are forced to fall back upon their "legal" personnel even for risky operations.

THE USE OF THE PARTY

The Communist party outside the Soviet Union has been used only intermittently by the Soviet government for actual espionage. Every time some element of the Communist party is caught in acts of espionage, this discredits the party as an "idealistic" and indigenous political organization and exposes it for what it really is—the instrument of a hostile foreign power, the stooge of Moscow. Whenever such exposures have taken place, as happened frequently in Europe in the 1920s, it has been observed that, for a time, there is a sharp decline in the intelligence work performed by local Communist parties. Furthermore, the value of using personnel not fully trained in intelligence work is questionable, since these amateur collaborators can expose not only themselves but also the operations of the intelligence service proper.

Chiefly in countries where the party is tolerated but where resident agents are difficult to procure, the Soviet intelligence services have had recourse to the party. This was the case in the United States during World War II. One of the reasons for the eventual collapse of Soviet networks that reached deeply into our government at that time was the fact that the personnel were not ideally suited for espionage. Many of these people had only strong ideological leanings toward Communism to recommend them for such work and in time were repelled by the discipline of espionage. Some, like Whittaker Chambers and Elizabeth Bentley, to whom the work became unpalatable, finally balked and volunteered their stories to the FBI. This problem came to a head for the Soviets just after the end of World War II as a result of the revelations of Igor Gouzenko, the defected code clerk of the Soviet embassy in Ottawa. At that time the KGB issued a secret order to its officers abroad not to involve members of Communist parties further in intelligence work.

The Communist party apparatus and Communist front organizations may, however, be useful for "spotting" potential agents for espionage. The evidence given in the Canadian trials by Gouzenko acquainted the public for the first time with the elaborate techniques employed by the Communist party under various guises. "Reading groups" and "study groups" for persons quite innocently interested in Russia were formed within Canadian defense industries, entirely for the purpose of spotting and cultivating people who could eventually be exploited for the information they possessed. The target in this case was the atomic bomb.

ENTRAPMENT

The Soviets often work on the principle that to get a man to do what you want, you try to catch or entrap him in something he would not like to have exposed to the public, to his wife, to his employers or to his government, as the case may be. If the potential victim has done nothing compromising, then he or she must be enticed into a situation set up by the KGB operatives which will be compromising. Two of the recent cases I have mentioned, that of Irvin Scarbeck in Poland and John Vassall in the Soviet Union, are examples of entrapment for intelligence purposes.

Within the Soviet Union itself, or in a bloc country, where the Soviets can set the stage, provide the facilities, a safe house, hotel or nightclub and furnish the cast of male or female *provocateurs*, tactics of entrapment are commonly used.

The sordid story of Vassall, the British Admiralty employee who spied for the Soviets for six years both in the Soviet Union and in London, is a typical one. In my own experience, I have run across a score of cases where the scenarios are almost identical with this one. The KGB operatives assigned to the task, after studying Vassall's case history from all angles and analyzing his weaknesses, set up the plan to frame him, exploiting the fact that he was a homosexual. The usual procedure here is to invite the victim to what appears to be a social affair; there the particular temptation to which the victim is likely to succumb is proffered him, and his behavior is recorded on tape or on film. He is then confronted with the evidence and told that unless he works for the Soviets the evidence will be brought to the attention of his employers. Vassall succumbed to this.

If the target individual is strong-willed enough to tell the whole story to his superior officer immediately, then the Soviet attempts at recruitment can be thwarted with relatively little danger to the individual concerned— even if he is residing in the Soviet Union. Sometimes his superior officer, particularly if the approach has been made in a free country, will want to play the man back against the Soviet apparatus in order to ferret out all the individuals and the tactics involved. Sometimes if the man approached does not seem qualified to play such a role, he is merely told to break off from his tormentors after telling them that he has disclosed everything.

The fact that the Soviets have no comeback when this is done is shown by an instance which came to light in the course of the official investigations into the Vassall case. The same Soviet agent employed in the British embassy in Moscow as a factotum who had originally drawn Vassall into a homosexual trap later attempted to recruit through blackmail a maintenance engineer of the embassy who had committed some black market offenses. The KGB expected that this victim, too, would rather cooperate with them than be exposed. The engineer, however, reported the recruitment attempt to his superiors, was promptly sent home from Moscow and the Soviet agent who had caused all the trouble finally lost his job with the British Embassy. At that time it was, of course, not known that he had also been responsible for the ploy which led up to the recruitment of Vassall.[3]

---

[3] It is possible that someone who has been or may be approached will see these lines; and this may help him to recognize the procedures. It can be hoped that he will take the path of full and frank disclosure advised here. If so, the ease with which the Soviet and sometimes the satellite operatives are able to effect recruitment will not be quite the same in the future.

Interestingly enough, we have found that some of the KGB operatives become so disgusted when forced to play the roles assigned to them in these recruitments that they become more willing candidates to break with it all and leave the service of the Soviet itself for a better life.

While homosexuality has played a prominent role in the most notorious recent cases, such as Vassall's, adultery or promiscuity is the more usual lever. Here, however, the Soviet and satellite intelligence services have learned over the years that blackmail based on the threatened exposure of illicit sexual acts is a powerful instrument when applied to men of certain nationalities, not so when applied to others. It depends on the mores, on the moral standards of the country of origin. The citizens of those countries where a certain value is placed on marital fidelity and where social disapproval of infidelity is strong are naturally the most likely victims.

I will refrain here from naming those countries which fall into the one category or the other in the opinion of the Soviets, since I would like to avoid opening an international debate on such a touchy subject. I cannot refrain, however, from passing along a story which was related to me some years ago at a time when the officials of a certain European satellite of the Soviets were still a little naïve about the attitudes in sexual matters of some of their Western neighbors. The secret police of the country in question had succeeded in taking some very compromising pictures of a certain diplomat which they hoped to use in order to force this gentleman to collaborate with their intelligence service. They invited him to their office under some pretext and showed him the pictures in their possession. They implied that the diplomat's wife as well as his superiors might be rather unhappy about him if they were shown the photographs. Contrary to their hopes and expectations, the diplomat didn't even wince at the implication but continued enthusiastically to study the photographs. Finally he said: "These are wonderful shots. I wonder if you fellows would be kind enough to make me some copies. I'd like two of these, and two of those. . . ." Either he was quite sophisticated or else he knew well how to handle blackmail.

An entirely different sort of pressure is that which the Soviets, as well as the satellites, bring to bear on refugees and expatriates who have close relatives behind the Iron Curtain. A refugee in the West may one day receive a visit from a stranger who will make the proposition clear to him: "Cooperate with us or your mother, brother, wife or children will suffer." However, since the refugee might just be courageous enough to complain to the local authorities, which could lead in turn to the apprehension of the agent who brought the message, the operation is more often run in less crude fashion. The refugee receives, instead of a visit, a letter from one of his close relatives at home which indicates in a veiled way that the local authorities are making inquiries about the refugee and that some unpleasantness may be in store for his relatives. This letter may be a forgery which the intelligence service has produced, especially if it is known that the refugee is not in frequent correspondence with any of his relatives. On the other hand, it may be authentic and the actual result of a visit which the police have paid to the relative. The refugee begins to worry and eventually writes a letter home asking how things are going. The relative, again under police direction or dictation, answers that things are going hard for them now but could be helped if the refugee would just do one or two little favors, one of these being to drop in at the embassy of his country for a chat. The intelligence service obviously gauges the likelihood of a successful recruitment by the tone of the letters the refugee writes back to his relatives and is not likely to risk the embarrassment of his exposing their tactics to authorities in the country of his adoption unless they see that he is falling for the game. Sometimes this technique is used to induce persons who have fled from Iron Curtain countries to return "home."

THE CHANGING PATTERN OF SOVIET OPERATIONS

The success of Soviet intelligence in the past and the depth of its penetration against its main targets are nowhere better evidenced than in its operations during World War II which have been uncovered. We must assume, however, that there were many such operations that have not come to light. Those that have are sufficient proof of an ability to establish and maintain clandestine contact with high-level sources under adverse conditions and to guide them in such a way that vital Soviet intelligence needs were fulfilled.

The key to many of these operations was the pro-Communist inclination of the people drawn into the networks and the important positions they occupied within their own governments or in sensitive installations. Klaus Fuchs, the atomic spy, is of course, a prime example of a case where the Soviets had an optimum intelligence advantage. Fuchs was employed in key British and American research installations and was a convinced Communist. Today, as we shall see, at least in the countries of its major opponents, the Soviets can no longer rely on finding such ideological collaborators in key positions. Hence they are forced more and more to turn to other tactics, chiefly entrapment or promises of sizable financial or other reward.

Soviet operations in World War II can be divided into two categories: those against its enemies and those against its "allies." In both areas Soviet intelligence had to fulfill Stalin's order "to get the documents," to reach directly into the places where decisions were made and literally to ferret out the facts and figures. The Soviets have never relied to the extent that Western countries have upon overt collection and expert analysis. Soviet intelligence, having developed within a highly secret and conspiratorial political atmosphere, naturally has an intense suspicion of the freely spoken or written word. The latter is of value to it only insofar as it serves to confirm or help interpret the intelligence produced by clandestine means—notably stolen documentary materials. In a country like Germany, even before the latter invaded Russia, and in Japan, with whom the Soviets were at peace until close to the war's end, it was the main aim of Soviet intelligence to find out what military preparations were being made which would affect the defense of the U.S.S.R.

In Japan, the major Soviet network run by the German Richard Sorge consisted almost entirely of Japanese officials and newspapermen close to the Cabinet, most of whom had been sympathizers with the Communist cause since their student days. The main achievement of the Sorge ring was to give Stalin by mid-1941 definite evidence that the Japanese then had no military intentions against the Soviets and were going to concentrate their forces against Southeast Asia and the Pacific—the Pearl Harbor tactic. This information was worth many divisions to Stalin, and he acknowledged his debt to Sorge but did nothing to save him once he was caught. Stalin was able to leave his eastern flanks only lightly fortified, confident that he would not have to fight on two fronts. The Sorge ring was rounded up shortly after this intelligence was received in Moscow, but it had done its job.

Against the Nazis and particularly the nerve centers of the German Army, Air Force and diplomatic service in Berlin, the Soviets ran a spy ring called the Schulze-Boysen-Harnack group. It was comparable to Sorge's ring in its makeup and mission. However, this group was by no means as professional in security techniques as Sorge's and was doomed to be found out sooner or later because of the carelessness of its members. It consisted of some thirty to forty anti-Nazi and pro-Communist sources scattered throughout Nazi ministries, the Armed Forces and the aristocracy.

Schulze-Boysen was an intelligence officer in the Air Ministry in Berlin. Harnack, whose wife, Mildred Fish, was an American (she and all of the ringleaders were executed), was an official in the Economics Ministry. The widely ramified contacts of these two men served the Soviets well. Of the hundreds of reports they passed in the period 1939–42, those of the greatest significance to the Soviets contained detailed information on the disposition of the German Air Force, German aircraft production, movements of ground troops and decisions of the German High Command—for example, the decision to encircle Leningrad and cut it off rather than attempt to occupy it.

The Gestapo unit that finally apprehended this group and other Soviet networks in Western Europe called them the *Rote Kapelle*, or Red Orchestra. After they were put out of operation by late 1942, the Soviets developed a fantastic source located in Switzerland, a certain Rudolf Rössler (code name, "Lucy"). By means which have not been ascertained to this day, Rössler in Switzerland was able to get intelligence from the German High Command in Berlin on a continuous basis, often less than twenty-four hours after its daily decisions concerning the Eastern front were made. Rössler was that unusual combination, a pro-Communist Catholic. Alexander Foote, who operated one of the secret Soviet radio bases that transmitted Lucy's information to Moscow, said of him:

> Lucy . . . held in his hands the threads which led back to the three main commands in Germany, and also could, and did, provide information from other German offices. . . . Anyone who has fought a battle from the general staff angle will know what it means to be able to place the flags of the enemy on the map and plan the disposition of one's own troops accordingly. . . . Lucy often put Moscow in this position, and the effect on the strategy of the Red Army and the ultimate defeat of the Wehrmacht was incalculable.[4]

---

[4] Alexander Foote, *Handbook for Spies* (New York: Doubleday & Co., Inc., 1949), p. 75.

The Sorge, *Rote Kapelle* and Lucy operations are the three best known of many Soviet penetrations in the war days. Altogether, the information which their intelligence work was able to collect through clandestine operations in World War II useful to the defense of the Soviets was about as good as any nation could hope to get.

In Allied countries the Soviet aim was essentially twofold. Stalin did not trust either Roosevelt or Churchill, and early in the game realized the coming clash of interests in the postwar world. Hence one aim of Soviet intelligence was to penetrate those offices of the American and British governments concerned with the "peace" settlements. The other targets were scientific and technological, in particular, nuclear. The Soviets knew that a great joint effort was being made in atomic research, and they wanted the benefits of it—hence Fuchs, Alan Nunn May, the Rosenbergs, Greenglass, Gold and a list of further names which came to light in the postwar years.

In the field of political intelligence, the cases and the agents have perhaps remained less fixed in the public memory, with the exceptions of the Hiss and Burgess-Maclean-Philby cases. The fact is, however, that in pursuit of their aim to learn what their ally the United Sates was planning for Germany, Central Europe and the Far East after the war, the Soviets had over forty high-level agents in various departments and agencies in Washington during World War II. At least this number was uncovered; we do not know how many remained undetected. Almost all of them, like the atomic spies, were persons of pro-Communist inclination at the time. Many have since recanted.

The Burgess-Maclean case, which broke in 1951 with the sudden flight of the two British officials to Soviet Russia, has perhaps been given too much the coloration of a defection. Also, its lurid angles have beclouded the real issues. This was no ordinary defection. The two men fled because they had timely warning from the "third man," Harold (Kim) Philby, that British security was hot on their trail. These three men in positions of trust in the British foreign service had been working for Soviet intelligence for years. All three were Communist sympathizers while students at Cambridge in the 1930s. They eventually became long-term Soviet penetration agents. Their value to the Soviets was increased as each served a tour of duty in the British embassy in Washington in the early 1950s. Philby's espionage activities were disclosed only in 1963, shortly after he had followed the other two behind the Iron Curtain.

In retrospect, it is Philby, less well known to the general public than his close friends, the notorious Burgess and Maclean, who deserves the closest scrutiny as perhaps the outstanding example of Soviet success in achieving high-level penetration through men who belonged to the generation of pro-Communist intellectuals of the twenties and thirties. Philby was not only a diplomat, useful as he and Burgess and Maclean may have been to the Soviets in this capacity; he was also a high-ranking intelligence officer.

In the postwar period, if we can judge from the cases that have been coming to light in the last ten years, Soviet intelligence in its pursuit of agents in sensitive positions in the U.S.A. and Britain began to run out of Communists and Communist sympathizers of the Fuchs-Rosenberg-Burgess-Maclean-Philby variety. There are a number of reasons for this. The hostile and aggressive intentions of Soviet Russia could no longer be masked by outwardly friendly diplomatic relations. The spectacle of the United States or Great Britain soft-pedaling a case of Soviet espionage because existing policy called for maintaining diplomacy on an even keel with the Soviets, a situation which prevailed from time to time in the late thirties and during the war, was unthinkable after about 1947. Instead, security precautions of a kind unprecedented in Western history began to be taken in our country and elsewhere to safeguard government offices, military establishments and sensitive scientific and industrial installations against penetration by employees who might be agents or potential Soviet agents. Secondly, the disillusionment with the once supposedly idealistic aims of Communism began to reach the intellectuals in the postwar period so that the late forties and fifties saw no groups of well-educated pro-Communists coming from the campuses of our universities and colleges, as had been the case from the depression days up to World War II.

The Soviets turned to other kinds of helpers, people who had other reasons for collaborating with them, willingly or unwillingly. Perhaps the most typical trend in the early postwar period, which illustrates the rapid adaptability of Soviet intelligence to new conditions, as well as the basic cold-blooded pragmatism of Communist tactics, was the massive recruitment by the Soviets of former SS and war criminals in both East and West Germany for intelligence work. The Soviets saw two strong factors they could exploit in dealing with such people. They were, first of all, by agreement of all the Allies, in the "automatic arrest" category. Under military government we had imprisoned many of them. The Soviets shot some of them. But what better way to force the recruitment of an agent than to stay his execution or excuse him from long imprisonment if he will consent to commit espionage in return for the favor? This was the line the Soviets took in East Germany. In West Germany, the de-Nazification procedures made it very difficult for former members of the SS, Gestapo and similar Nazi organizations to get decent jobs. Many of these men who had shortly before been riding high under the emblems of Nazi power were ostracized, unemployed and in dire straits. Their attitude toward the American and British occupation authorities was, to say the least, negative. They were ripe for the Soviet invitation to treason. They hardly felt it to be treason, since in their opinion there was with Germany under foreign military rule no real authority to which they felt any direct loyalty.

A case of this kind was that of Heniz Felfe, a senior officer of the West German intelligence service, who was caught by his own colleagues and superiors in November, 1961, after having betrayed what he knew of their work to the Soviets ever since he had joined the service over ten years before. In 1945 Felfe had been a rather junior member of the foreign arm of the Nazi security and intelligence service. He hailed from a part of Germany which came under Soviet occupation after the war was over. He had been captured and interned in Holland by the Allies and after his release tried to settle in West Germany. He went through the de-Nazification process but had great difficulties finding a job to his liking. Eventually, armed with questionable credentials and letters of recommendation he had talked some innocent people into giving him, he applied for a police job, the only kind of work he knew. In the rather confused atmosphere of the Allied-sponsored German civil service, he got a job in a minor office of the counterintelligence section. Later it turned out he had been helped to the job by certain German officials who themselves were under Soviet pressure.

During this period, Felfe himself became a Soviet agent, having fallen into Soviet clutches while on a secret trip to his home area of East Germany. The man who led the Soviets to him was a friend, also a former SS man, who had made his bargain with the Soviets at an even earlier date. Felfe, in turn, recommended others of similar ilk. The price of all this was cheap for the Soviets—past sins were forgiven and a little money and protection were offered for the future. But a sword hung over the heads of these people, and they knew it would fall if they betrayed the Soviets. The Soviets picked up all the old SS men they could find. Most of them were guaranteed to be ambitious and utterly unprincipled. A few would be clever enough to work their way up the ladder of the West

German civil service. Felfe was one of these, and the Soviet investment paid off handsomely.

The case of Felfe was one of Soviet recruitment based on a Nazi past. The KGB, however, is just as ready to use old and hidden Communist connections where the victim to be recruited is working in the West and where his future is dependent upon creating the impression that he has had nothing to do with Communism. Such were the facts in the important case of Alfred Frenzel, a prominent member of the West German parliament (Bundestag), to which he was first elected in 1953. For some years he served on the parliamentary committee which dealt with matters of German defense, and in this capacity he had access to information relating to the build-up and equipment of the West German military forces and NATO plans therefor.

Frenzel had originally come from the Sudetenland of Czechoslovakia. There for a time he had been a member of the Communist party; in fact had been thrown out of the party under the accusation of embezzling party funds. All this was well known to the Czechoslovak secret service.

Frenzel, like so many of his fellow Sudeten Germans, became a refugee in West Germany in the postwar days. He entered politics there, had considerable success and felt that he had securely buried the past. When the Czechoslovak secret service approached him in the mid-1950s and threatened to ruin his career with a full disclosure of his Communist affiliations unless he cooperated, Frenzel easily succumbed. He was an ideal "set-up" for recruitment, a man in a prominent and sensitive political position with a secret and rather lurid Communist past: disclosure spelled ruin for him. Here, as in the Felfe case, the Soviets could offer him financial help and protection. For some five years prior to October, 1960, when he was arrested, he had been working for the Czechoslovak secret service and, through them, for the Soviets; and his intelligence masters saw to it that he produced "the goods" to compensate for the protection and favors granted.

There were also several cases of recruitment in West Germany based upon evidence that the victims had had abortions in the Eastern zone before fleeing westward. This vulnerability was carefully tabulated and used. It was thus that Rosalie Kunze, the secretary of Admiral Wagner, Deputy Chief of the German Navy, was recruited by the Soviets. In some cases, doctors who in their East German past had committed illegal operations were followed and became targets for recruitment when they came to West Germany.

But such displaced rootless vagrants of postwar Europe are only one type of agent that Soviet intelligence is looking for. Among those who still have home and country the Soviets will search out the misfits and the disgruntled, people in trouble, people with grievances and frustrated ambitions, with unhappy domestic lives—neurotics, homosexuals and alcoholics. Such people sometimes need only a slight nudge, a slight inducement to fall into the practice of treason. Sometimes entrapment is necessary, sometimes not.

The Soviets are, of course, well aware of the fact that persons with moral and psychological weaknesses do not make the best agents. They only use them where there is nothing better available. They would prefer the ideologically motivated people and still keep on the lookout for them.

If the postwar world presented the Soviets with a somewhat different breed of spies from the ideological types they had concentrated on in earlier years, it also presented them with brand new targets—NATO, for example. For a time at least, this was perhaps the most important target, representing as it does a powerful coalition of forces the Soviet considers potentially hostile. The lure of NATO's structure from the point of view of Soviet intelligence is the access all its members have to important military secrets of the major participants. It is not necessary to recruit an American to get at American secrets we share with NATO. At the same time, of course, the overall plans of NATO itself are of prime importance to the Soviets. A Belgian, Frenchman or German serving with NATO can get his hands on both kinds of secrets.

On July 7, 1964, the Frenchman, Georges Paques, who had been deputy press chief at NATO headquarters in Paris before his arrest in 1963, was sentenced to life imprisonment for treason to France. Since NATO itself is neither a sovereign entity nor a judicial body, a man cannot be tried by NATO itself or condemned for being treasonable to it. But the fact is that Paques did a great deal of damage to France, America and NATO by passing documents, chiefly of a political-military nature, concerning all three to the Soviets, including, it is reported, Allied military contingency plans for Berlin, NATO force goals and other NATO military matters. He stated before the court that he did so in order to avert war, "to assure France's survival" and "to try to save mankind."

He professed to detest everything American and saw NATO as an "American dominated institution." He claimed that there was nothing treasonable to France in betraying American secrets to the Soviets. The French court did not accept this cleverly contrived excuse and furthermore felt he had also betrayed enough French secrets to deserve a heavy penalty. The prosecutor actually asked for the death penalty, but the court gave a life sentence. Paques' subtle defense was in all likelihood a divisive tactic suggested by the Soviets themselves. It made him appear quasi-innocent in the eyes of some people in France. Also the appeal to anti-American sentiments was secretly pleasing to some French quarters.

An interesting point in the Paques case is that the man does not appear to have been a Communist, although his wife and some of her relatives were at one time. Politically Paques himself was known as a rightist of a rather extreme sort. Of course, this may have been a protective pose. In any case, there is no history here of an earlier intellectual flirtation with Marxism. This was not because of any intellectual shortcomings on Paques' part. On the contrary, he was an intellectual snob who looked down on the mental capabilities of some of the Soviet spymasters with whom he dealt over the years. The most recent of these, Vasily Vlasov, First Secretary of the Soviet embassy in Paris, was apparently regarded by Paques as his intellectual equal, and the Soviet benefited accordingly from Paques' cooperation. This illustrates the point that an intelligence service can get more out of an agent by putting someone next to him who is in tune with him and whom he can respect intellectually. The Soviets apparently put up with Paques' intellectual vanity, since his contributions to their knowledge made him more than "tolerable," to say the least.

A more exclusively military than political case was that of the Swedish Colonel Stig Wennerstrom, who was sentenced to life imprisonment by a Swedish tribunal in May, 1964. Here again, as in the Paques case, the betrayal was threefold. Wennerstrom passed to the Soviets some Swedish, American and even NATO military secrets, which came into Swedish hands even though Sweden was not a member of NATO.

In the course of this service to the Soviets he was secretly made a Soviet citizen and promoted to the rank of major general in the Soviet army (though he rose no higher than a colonel in the service of his own country). It is a rather interesting trick of the Soviets, which costs them nothing, that among other forms of payment, they bestow upon their best agents not only Soviet citizenship (which may be taken up if the agent is forced to flee to the Soviet Union or goes there to spend his retirement years) but also a military rank, a calculated piece of flattery which no doubt appears to frustrated opportunists like Wennerstrom to be a tangible reward even though they may never get a chance to wear the uniform that goes with the rank—at least not in public.

Wennerstrom also accrued a tidy fortune from the Soviets, much of which was put aside for him in Russia for later use. Probably the Soviet feared that the temptation to him to use the money might be too great to resist and that heavy spending would give him away. He will, as things now stand, never have the pleasure of spending his Soviet hoard.

On a somewhat lower plane, there was the case in Iceland recently of two Soviet diplomats who were expelled because they tried to pressure a young Icelandic trucker into committing espionage for the Soviet Union. They wanted him to get information for them on the NATO Air Base at Keflavik. What makes the case interesting and symptomatic of the changed times is the fact that the victim, a certain Ragnar Gunnarsson, a man of thirty-two, was a card-carrying Communist and still is— at least he still was in February, 1963.

Yet it was this Communist who refused to submit to Soviet pressure and who informed the Icelandic police of the whole plot and even cooperated with them in trapping the Soviets in the act.

The Soviets had cultivated Gunnarsson for a long time. When he was only twenty-two, he had been invited to the Soviet Union for a three-week tour with eight other Icelandic youths and had been shown the sights at Soviet expense. Later the Soviets tried to cash in on the investment, but they picked the wrong man or, what is more likely, they had yet to learn that times have changed. It is possible now for a Communist not to feel obliged to spy for the Soviet Union and even to take steps to frustrate their espionage. Whittaker Chambers and Elizabeth Bentley went to the FBI in 1945 and revealed what Soviet espionage was doing in the United States *after* they had been involved in it themselves for years. By then they were entirely disillusioned and broke with Communism entirely. Gunnarsson refused to commit espionage in the first place, but remained a Communist.

What apparently makes such a state of mind as Gunnarsson's possible today is the fact that the Soviet Union is no longer the holy matrix of Communism (in the eyes of its adherents), but only a sponsor of it, and one of several sponsors at that. And this seems to have set back the Soviet intelligence services in their search for agents. The ground has been taken away from under the ideological appeal to commit espionage in all but the backward countries.

The case which was exposed in Australia in February, 1963, points more sharply than any other to the failure of the vaunted Soviet service to keep up with a changing world and to manage its business successfully among strangers and in a country where good security practices prevail. The Soviets had suffered an enormous setback in Australia in 1954 when the KGB resident, Vladimir Petrov, defected. One reason he defected was because he saw even at that time that the tasks the KGB had assigned him in Australia were hopeless, that the KGB in Moscow could not understand that Australia in 1954 was not, let us say, like Germany in the late 1920s. And he knew that he himself would be blamed for Moscow's failure to adjust to a new situation.

His defection and his disclosures of Soviet espionage in Australia caused a break in diplomatic relations between the Soviet Union and Australia which were only resumed again in 1959. By this time there was an attempted "new look" to Soviet espionage tactics noticeable in many places. The very man who was sent to head up the reopened Soviet Embassy in Canberra, Ivan Skrypov, was a high KGB official under diplomatic cover, evidence that the espionage task had first priority in Soviet eyes. After all, there was lost time to be made up for. But Skrypov was not the sinister, silent type of the old school. He was a gay blade, a party-giver, a backslapper. His open participation in Australian official life was supposed to mislead everyone as to his true mission. This "new look" also was apparent in the social cavortings of Captain Yevgeni Ivanov, Soviet Deputy Naval Attaché and intelligence officer in London during the early 1960s, who allegedly shared the favors of Christine Keeler with the British Minister of War, John Profumo.

Behind the backs of his genial Australian hosts, Skrypov was going about his real job—to build up a new undercover intelligence apparatus in Australia. In the pursuit of his task he made, however, one serious error. He hired for certain specialized functions an Australian woman who was really an agent of the Australian Security Service. This was the kind of coup on the part of the Australians that the Soviets themselves have tried to practice so often, yet it has rarely been practiced successfully against them, largely because in the past they did not have to rely on strangers and outsiders, and when they did, their own investigative capabilities could usually determine how reliable the agent was, i.e., they tailed him around and checked him out. Here, in a strange land with a strong and watchful security service, however, the Soviets could neither pick up local Communist sympathizers for their work nor could they muster enough "leg-men" and informers to keep track of their main agents. Thus they had to rely on the show of "goodwill" and apparent dedication of their "volunteer." Their ability to judge behavior was hampered because they were dealing with a species of people foreign to them.

The blow to the Soviets in Australia was well deserved. What Skrypov was trying to do through his agent was to set up an illegal *residentura* for the KGB which would have obviated use of the Soviet embassy for important espionage operations. Thus a high-speed radio transmitter and other materials for clandestine work were passed to the agent for a further party in Adelaide who was later to function illegally. In apprehending Skrypov through their double agent, the Australians put both the legal and illegal apparatus of the KGB in Australia out of business for a long time to come. Whether the Soviets will try a third time to create an espionage apparatus in Australia remains to be seen.

Without wishing to appear overly optimistic, I would hazard the guess that the KGB will for the moment retreat, mete out the appropriate punishments to the officers at fault in this latest fiasco and wait a time before trying again. Then they will probably come up with some entirely new scheme for penetrating the Australian defenses. They will certainly "case the joint" more carefully in the future. What they may realize, though they may never give up, is that in a country which is aware and knowledgeable of Soviet aims and tactics and is willing to make a serious effort to guard itself by maintaining a highly trained, competent security and counterintelligence force, success for the Soviet spy is difficult. This is particularly true of a country like Australia, where indigenous Communism is feeble.

Following the exposure and expulsion of Skrypov by the Australians, the Soviets retaliated, as they often do, by looking around for some way in which they could embarrass the Australians. In general, whenever a Western power catches and expels a Soviet diplomat engaged in espionage or other illegal activities, the Soviets will select a diplomatic representative of that same power in Moscow, more or less at random, although he must be of suitable rank, and declare him *persona non grata*. This puts a certain strain on the West, since an adequate replacement must be found.

In the Australian case, the Soviet practice took a rather ludicrous turn. Within the relatively small Australian embassy in Moscow, it was difficult to find a ranking member on whom any trumped-up story could be hung with even minimal credibility. The Soviets eventually selected First Secretary William Morrison, declared him *persona non grata* and charged him with collecting intelligence and illegally selling foreign clothes to Soviet citizens. This last bit shows that the intelligence charge was so weak the Soviets evidently felt it necessary to tack on an additional complaint just to cover themselves. That a foreign diplomat would engage in the vending of secondhand garments is about as ludicrous a charge as one can imagine. Unfortunately, under diplomatic procedures there is no recourse or appeal when one country declares the diplomat of another *persona non grata*. Hence this practice is subject to abuse and to exercise by way of retaliation without either rhyme or reason.

If illegals or other agents without diplomatic status are caught and sentenced for espionage, then quite another reciprocal procedure may take place between the Soviets and the Western powers—the exchange of prisoners. The most striking example of this was the exchange in February, 1962, of Francis Gary Powers and another American, Frederic Pryor, held in the Soviet Union on charges of espionage, for the Soviet spy Colonel Rudolf Abel. This had several interesting implications. First of all, it meant the breakdown of Soviet pretensions that they had no responsibility for Abel, a position they took at the time of his arrest, trial and conviction; and secondly, it opened up the possibility that the exchange of spy for spy might become a general practice. I was Director of Central Intelligence when the secret negotiations for the Powers-Abel exchange were initiated, and I approved of them. While I had some misgivings, on the whole, I felt then and feel now that it was a fair exchange and that it was in our own interest to proceed with it under the particular circumstances of this rather unusual case. However, this has tended to create a precedent which may have some unfortunate consequences. The number of Soviet agents in the West, we may assume, greatly exceeds the number of Western agents behind the Iron Curtain. Hence with reasonable competence and vigilance on our part, we are likely at any given time to have in our control more Soviet agents than the number of Western agents that they are detaining. If the idea of swapping agent for agent becomes the practice, the Soviet will be anxious to have a backlog of apprehended agents in their hands. Hence they will be tempted, and will likely succumb to the temptation, to arrest casual visiting Westerners who have nothing whatever to do with intelligence.

In the early summer of 1963 it was rumored that another exchange of captured agents was under consideration. In the last two years the British succeeded in apprehending, convicting and imprisoning seven major Soviet agents, Blake, Vassall and the five members of the Lonsdale ring: Lonsdale himself, Houghton and his girl friend, and the Kroger pair. During the same period the Soviets caught and imprisoned only one Britisher on an espionage charge. This was Greville Wynne, the London businessman whom the Soviets accused of serving as an intermediary to Oleg Penkovsky, since executed. Wynne received eight years from the Soviet court. The combined prison sentences of the seven persons in British hands amounts to something over 150 years. The bargaining position of the Soviets is obviously not a strong one. The man they most wanted to see released was obviously Lonsdale because he is the only one of the seven who is a Soviet national and, like Abel, he is a long-term illegal. Rumor has it, however, that the Soviets are also interested in freeing the Krogers, who undoubtedly have served them well for decades.[5]

Before we go much farther down this road of swapping spies, it would be well to have a look and see where it may lead.

In mid-October of 1963, two American prisoners of the Soviets, Walter Ciszek, a Catholic priest who had been in Soviet captivity for twenty-three yeas, and Marvin Makinen, a young student, were exchanged for two Soviet espionage agents picked up in the United States by the FBI in August, 1963. In this exchange it would appear that the Russians gave up nothing of value to themselves but realized a very significant gain in recovering two well-trained and experienced operatives. With the release of Ciszek and Makinen, however, the Soviets evidently scraped the bottom of the barrel, and the trumped-up case against Professor Barghoorn, which followed shortly after, may well have been nothing but a bare-faced attempt to seize a fresh hostage. Professor Barghoorn, arrested by the Soviets on the streets of Moscow in November of 1963, who was quite innocent of any charge of espionage, would quite likely have been held by the Soviets as a pawn of highest value in reserve against the exchange of Soviet agents we might apprehend in the future. However, this incident backfired in the faces of the Soviet policymakers, thanks to President Kennedy's vigorous action.

[5] The Soviets have, in fact, succeeded in exchanging Wynne for Lonsdale. The exchange took place on April 22, 1964, at a West Berlin border point. The British, knowing the trade was an uneven one, allegedly acceded to it out of humanitarian motives because Wynne was reported ill in his Soviet jail.

8

Counterintelligence

In today's spy-conscious world, each side tries to make the opponent's acquisition of intelligence as difficult as possible by taking "security measures" in order to protect classified information, vital installations and personnel from enemy penetration. These measures, while indispensable as basic safeguards, become in the end a challenge to the opponent's intelligence technicians to devise even more ingenious ways of getting around the obstacles.

Clearly, if a country wishes to protect itself against the unceasing encroachments of hostile intelligence services, it must do more than keep an eye on foreign travelers crossing its borders, more than placing guards around its "sensitive" areas, more than checking on the loyalty of its employees in sensitive positions. It must also find out what the intelligence services of hostile countries are after, how they are proceeding and what kind of people they are using as agents and who they are.

Operations having this distinct aim belong to the field of counterespionage, and the information that is derived from them is called counterintelligence. Counterespionage is inherently a protective and defensive operation. Its primary purpose is to thwart espionage against one's country, but it may also be extremely useful in uncovering hostile penetration and subversive plots against other free countries. Given the nature of Communist aims, counterespionage on our side is directly concerned with uncovering secret aggression, subversion and sabotage. Although such information is not, like positive intelligence, of primary use to the government in the formation of policy, it often alerts our government to the nature of the thrusts of its opponents and the area in which political action on our part may be required.

In 1954, the discovery of concealed arms shipments, a whole boat-load of them, en route from Czechoslovakia to Guatemala first alerted us to the fact that massive Soviet support was being given to strengthen the position of a Communist regime in that country.

The function of counterespionage is assigned to various U.S. agencies, each of which has a special area of responsibility. The FBI's province is the territory of the United States itself, where, among other duties, it guards against the hostile activities of foreign agents on our own soil. The CIA has the major responsibility for counterespionage outside the United States, thereby constituting a forward line of defense against foreign espionage. It attempts to detect the operations of hostile intelligence before the agents reach their targets. Each branch of the armed forces also has a counterintelligence arm whose purpose is mainly to protect its commands, technical establishments and personnel both at home and abroad against enemy penetration.

The effectiveness of this division of labor depends upon the coordination of the separate agencies and on the rapid dissemination of counterintelligence information from one to the other.

It was a coordinated effort that resulted in the capture of Soviet spymaster Colonel Rudolf Abel. In May, 1957, Reino Hayhanen, a close associate and co-worker of Colonel Abel in the United States, was on his way back to the Soviet Union to make his report. While in Western Europe, he decided to defect and approached U.S. intelligence, showing an American passport obtained on the basis of a false birth certificate. Hayhanen's fantastic story of espionage included specifics as to secret caches of funds, communications among agents in his network and certain details regarding Colonel Abel. All this information was immediately transmitted to Washington and passed to the FBI for verification. Hayhanen's story stood up in every respect. He came back willingly to the United States and became the chief witness at the trial against Abel.

As soon as Hayhanen reached our shores, primary responsibility for him was transferred to the FBI, while CIA continued to handle foreign angles.

The classical aims of counterespionage are "to locate, identify and neutralize" the opposition. "Neutralizing" can take many forms. Within the United States an apprehended spy can be prosecuted under the law; so can a foreign intelligence officer who is caught red-handed if he does not have diplomatic immunity. If he has immunity, he is generally expelled. But there are other ways of neutralizing the hostile agent, and one of the best is exposure or the threat of exposure. A spy is not of much further use once his name, face and story are in the papers.

The target of U.S. counterespionage is massive and diverse because the Soviets use not only their own intelligence apparatus against us, but also those of Poland, Czechoslovakia, Hungary, Rumania and Bulgaria, all of which are old in the ways of espionage if not of Communism. Chinese Communist espionage and counterespionage operations are largely independent of Moscow, though many of their senior personnel in earlier days were schooled by Soviet intelligence.

Although the purpose of counterespionage is defensive, its methods are essentially offensive. Its idea goal is to discover hostile intelligence plans in their earliest stages rather than after they have begun to do their damage. To do this, it tries to penetrate the inner circle of hostile services at the highest possible level where the plans are made and the agents selected and trained, and, if the job can be managed, to bring over to its side "insiders" from the other camp.

One of the most famous cases of successful high-level penetration of an intelligence service is that of Alfred Redl, who from 1901 to 1905 was chief of counterespionage in the Austro-Hungarian Empire's military intelligence service, and later its representative in Prague. From the available evidence it would appear that from 1902 until he was caught in 1913 Redl was a secret agent of the Russians, having been trapped by them early in his intelligence career on the basis of two weaknesses—homosexuality and overwhelming venality. He also sold some of his wares at the same time to the Italians and the French. But that wasn't all. As a leading officer of the military intelligence, Redl was a member of the General Staff of the Austro-Hungarian Army and had access to the General Staff's war plans, which he also gave to the Russians.

Despite the fact that Redl was apprehended just before the war, his suicide at the "invitation" of his superior officers immediately after his treachery was discovered eliminated the possibility of interrogating him and determining the extent of the damage he had done. The Austrians were more interested in hushing up the scandal. Even the Emperor was not told of it at first.

Ironically enough, Redl was caught by a counterespionage measure—postal censorship—which he himself had developed to a point of high efficiency when he had been counterespionage chief. Two letters containing large sums of banknotes and nothing else were inspected at the General Delivery Office of the Vienna Post Office. Since they had been sent from a border town in East Prussia to a most peculiar-sounding addressee, they were considered highly suspicious. For almost three months the Austrian police doggedly waited for someone to come and collect the envelopes. Finally Redl came, and the rest is history. However, it still amazes counterintelligence specialists who study the case today that the Russians, in an operation of such immense significance to them, could have resorted to such careless devices for getting money to their agent, especially since postal censorship was one of the favorite counterespionage devices of the Okhrana itself.

It is, of course, not necessary to recruit the chief, as in the Redl case. His secretary, had he had one, might have done almost as well. Actually, the size of a major intelligence organization today makes it impossible for the chief to be concerned with all the operational details an opposing service would wish to know. Not only that, but today the headquarters of an intelligence organization are as "impenetrable" as the best minds assigned to the task can make them. As a consequence, counterespionage usually aims at more accessible and vulnerable targets directly concerned with field operations. These targets will often be the offices and units which intelligence services maintain in foreign countries. As is well known, they are frequently found in embassies, consulates and trade delegations, which may afford the intelligence officer the protection of diplomatic immunity as well as a certain amount of "cover."

How does the counterespionage agent "penetrate" his target? By what means can he gain access to the personnel of another intelligence service? One of the ways is to come supplied with beguiling information and offer it and his services to the opposition. Since some of the most crucial intelligence in recent history has been delivered by people who just turned up out of a clear sky, no intelligence service can afford to reject out of hand an offer of information. Of course, behind the Iron Curtain and in most diplomatic establishments of the Soviet bloc outside the Curtain, the general distrust and suspicion of strangers is such that an uninvited visitor, no matter what he is offering, may not go beyond the receptionist.

In the end, however, his ability to get a foot in the door depends on the apparent quality of the information he is offering. Every intelligence service has the problem of distinguishing, when such unsolicited offers come along, between a bona fide volunteer and a penetration agent who has been sent in by the other side. This is no easy matter.

If counterespionage succeeds in "planting" its penetration agent with the opposing service, it is hoped that the agent, once he is hired by the opposition, will be given increasingly sensitive assignments. All of them are reported duly by the agent to the intelligence service running the "penetration."

The Soviets used this method against Allied intelligence offices in West Germany and Austria during the 1950s. Refugees from the East were so numerous at that time that it was necessary to employ the better-educated ones to help in the screening and interrogation of their fellow refugees. The Soviets determined to take advantage of this situation and cleverly inserted agents in the refugee channel, providing them with information about conditions behind the Curtain which could not fail to make them seem of great interest to Western intelligence. Their task for the Soviets was to find out about our methods of handling refugees, to get acquainted with our personnel and also to keep tabs on those among the refugees who might be susceptible to recruitment as future Soviet agents.

This same penetration tactic can be used to quite a different end, namely, provocation, which has an ancient and dishonorable tradition. The expression *agent provocateur* points to French origins and was a device used in France during times of political unrest, but it is the Russians again who made a fine art of provocation. It was the main technique of the czarist Okhrana in smoking out revolutionaries and dissenters. An agent joined a subversive group and not only spied and reported on it to the police, but incited it to take some kind of action which would provide the pretext for arresting any or all of its members. Since the agent reported to the police exactly when and where the action was going to take place, the police had no problems.

Actually, such operations could become immensely subtle, complicated and dramatic. The more infamous of the czarist *agents provocateurs* have all the earmarks of characters out of Dostoevski. In order to incite a revolutionary group to the action that would bring the police down on it, the *provocateur* himself had to play the role of revolutionary leader and terrorist. If the police wished to round up large numbers of persons on serious charges, then the revolutionary group had to do something extreme, something more serious than merely holding clandestine meetings. As a result, we encounter some astounding situations in the Russia of the early 1900s.

The most notorious of all czarist *provocateurs*, the agent Azeff, appears to have originated the idea of murdering the Czar's uncle, the Grand Duke Sergius, and the Minister of the Interior, Plehwe. The murders then gave the Okhrana the opportunity of arresting the terrorists.

One of Lenin's closest associates from 1912 until the Revolution, Roman Malinovsky, was, in fact, a czarist police agent and *provocateur*, suspected by Lenin's entourage but always defended by Lenin. Malinvosky helped reveal the whereabouts of secret printing presses, secret meetings and conspiracies to the police, but his main achievement was far more dramatic. He got himself elected, with police assistance and with Lenin's innocent blessing, as representative of the Bolshevik faction to the Russian parliament, the Duma. There he distinguished himself as an orator for the Bolsheviks. The police often had to ask him to restrain the revolutionary ardor of his speeches. Indeed, in the cases of both Azeff and Malinovsky, as with many "doubles," there is some question as to where their allegiance really lay. Since they played their "cover" roles so well, they seem at times to have been carried away by them and to have believed in them at least temporarily.

Nowadays when you read in the paper that an individual has been expelled from one of the Soviet bloc countries, it is frequently either a completely arbitrary charge, often in reprisal for our having caught and expelled a Soviet bloc intelligence officer in the Untied States, or else it is the result of a provocation.

The routine goes like this. One day a foreigner behind the Iron Curtain is called upon at home or encountered in a restaurant, on the street or even in his office by a member of the "underground" or by someone who feigns dissatisfaction with the regime and offers important information. The "target" may accept the information and continue to meet the informant. If so, sooner or later during one of these meetings, the local security police "arrest" the informant for giving information to a foreign power. The target may find his name in the paper, and, if he is an official, his embassy will receive a request from the local Foreign Office that he leave the country within twenty-four hours. The informant was, of course, a provocation agent planted by the police.

Even though these incidents are generally faked, much of the world audience whom the Soviets try to impress will not recognize them for what they are. Whenever the Soviets can accuse the West of spying, of abusing their diplomatic privileges, of meddling in the affairs of the "peace-loving socialist republics," they will do so; and concrete instances of Westerners "caught in the act" provide the best ammunition for their propaganda.

The double agent is the most characteristic tool of counterespionage operations, and he comes in many guises. In an area like West Germany, with its concentration of technical and military installations, both those of the West Germans and of the NATO forces, there is a flood of agents from the Soviet bloc spying on airfields, supply depots, factories, United States Army posts, etc. Many are caught. Many give themselves up because they have found a girl and want to say with her or simply because they find life in the West more attractive. Such men become double agents when they can be persuaded to keep up the pretense of working for the Soviet bloc under Western "control." The ones who are caught often agree to this arrangement because it is preferable to sitting in jail for a couple of years.

The aim is to build up the agent, allowing him to report back to the bloc harmless information, which is first screened. It is hoped that the Soviets will then give him new briefs and directives, which show us what the opponent wants to know and how he is going about getting it. Sometimes it is possible, through such an agent, to lure a courier or another agent or even an intelligence officer into the West. When this happens, one has the choice of simply watching the movements of the visitor, hoping he will lead to other agents concealed in the West, or of arresting him, in which case the operation is naturally over, but has succeeded in neutralizing another person working for the opposition.

A more valuable double is the resident of a Western country who, when approached by an opposition intelligence service to undertake a mission for them, quietly reports this to his own authorities. The advantages are obvious. If the Soviets, for example, try to recruit a Westerner, they must have something serious in mind. Secondly, the voluntary act of the person approached, in reporting this event, points to his trustworthiness. The target of Soviet recruitment will usually be told by his own intelligence authorities to "accept" the Soviet offer and to feign cooperation, meanwhile reporting back on all the activities the Soviets assign him. He is also provided with information which his principals desire to have "fed" to the Soviets. This game can then be played until the Soviets begin to suspect their "agent" or until the agent can no longer stand the strain.

The case of the late Boris Morros, the Hollywood director, was of this kind. Through Morros, who cooperated with the FBI for many years, the Soviets ran a network of extremely important agents in the United States, most of them in political and intellectual circles. This operation led to the apprehension of the Sobles, of Dr. Robert Soblen and numerous others.

"Surveillance" is the professional word for shadowing or tailing. Like every act of counterespionage, it must be executed with maximum care lest its target become aware of it. A criminal who feels or knows he is being followed has limited possibilities open to him. The best he can hope for is to elude surveillance long enough to find a good hiding place. But an intelligence agent, once he has been alarmed by surveillance, will take steps to leave the country, and he will have plenty of assistance in doing so.

The purpose of surveillance in counterespionage is twofold. If a person is only suspected of being an enemy agent, close observation of his actions over a period of time may lead to further facts that confirm the suspicion and supply details about the agent's mission and how he is carrying it out. Secondly, an agent is rarely entirely on his own. Eventually he will get in touch, by one means or another, with his helpers, his sources and perhaps the people from whom he is taking orders. Surveillance at its best will uncover the network to which he belongs and the channels through which he reports.

Surveillance was largely responsible for the British success in rounding up five Soviet agents in the Lonsdale ring in January, 1961. Harry Houghton, an Admiralty employee, was suspected of passing classified information to an unidentified foreign power. Scotland Yard tailed Houghton to a London street, where he met another man so briefly that it was impossible to tell for certain whether anything had passed between them or whether they had even spoken.

However, the fact that both parties acted furtively and seemed extremely wary of surveillance convinced the British that they were on the right track. The Yard split is trained men into two teams to follow the suspects separately. This eventually led them, after many days of tireless and well-concealed surveillance, to a harmless-looking American couple who operated a secondhand book store. Their role, if any, could not be immediately ascertained.

On a later occasion Houghton came up to London again, this time with his girl friend, who worked in the same naval establishment. Again under surveillance, the two of them, walking down the street carrying a market bag, were approached from the rear by the same man whom Houghton had met previously. Just as this fellow was about to relieve Houghton and the girl of the market bag, which was clearly a pre-arranged method for passing the "goods," all three were arrested. The unknown man was Gordon Lonsdale, the Soviet "illegal" with Canadian papers who was running the show.

A few hours later, the harmless-looking American booksellers met the same fate. They were being sought by the FBI for their part in a Soviet net in the United States and had disappeared when things had become too hot for them. In London they had been operating a secret transmitter to relay Lonsdale's information to Moscow.

Counterintelligence, like most branches of intelligence work, has many technical resources, and one among them has been responsible in the past for uncovering more concealed intelligence networks than any other single measure. This is the interception and locating of illegal radio transmitters, known as "direction-finding," or D/Fing for short. It employs sensitive electronic measuring devices which, when mounted on mobile receivers, in a car or truck, can track down the location of a radio signal by indicating whether the signal is getting stronger or weaker as a mobile receiver weaves around a city listening to what has already been identified as an illegal transmitter.

Every legal radio transmitter, commercial or amateur, in most countries today is licensed and registered. In this country the call signal and the exact location of the transmitter are on record with the Federal Communications Commission. The FCC monitors the air waves at all times as a law-enforcement procedure. This leads to the uncovering of enthusiastic "ham" radio operators who haven't bothered to get a license. It also leads to the discovery of illegal agent transmitters. The latter are usually identifiable because their messages are enciphered and they do not use any call signal on record.

Monitoring of a suspicious signal may also reveal that the operator has some kind of fixed schedule for going on the air, and this almost unfailingly points to the fact that he is transmitting to a foreign headquarters by prearrangement. At this point the D/Fing process begins. The main difficulty of tracking is that the illegal operator usually stays on the air, for obvious reasons, only for very short periods. As the mobile D/F experts try to trace his signal across a large city on air waves crowded with other signals, he suddenly finishes, goes off the air, and there is nothing the D/Fers can do until he comes on again some days or weeks later. If the Soviets are behind the operation, the transmission schedule, while fixed, may follow a pattern that is not easy to spot. Also, the transmitting frequency may change from time to time. The only solution is for the D/F headquarters to listen for the suspicious signal all the time and to keep after it. But here, too, the technicians have invented new improvements to foil and outwit each other. The latest is a high-speed method of transmission. The operator does not sit at his telegraph key sending as fast as he can. He prerecords his message on tape, then plays the tape over the air at breakneck speed, too fast for any ear to disentangle. His receiving station at home records the transmission and can replay it at a tempo which is intelligible. If the illegal operator is on the air for only twenty or thirty seconds, the D/Fers are not going to get very far in their attempt to pinpoint the physical location of the transmitter.

During World War II, before the invention of these high-speed techniques, the efficiency of D/Fing on both sides was responsible for some very dramatic counterintelligence work. In the famous Operation Northpole, British intelligence headquarters in London was in touch with the Dutch underground by radio. The Dutch center radioed intelligence on German military matters to London and also made arrangements by wireless with London to have further personnel and equipment air-dropped into Holland. From 1942 to 1944 the British, complying with the requests and arrangements proposed by the various Dutch underground radio transmitters, dropped large amounts of weapons and supplies into Holland at prearranged drop areas. Many of the bombers which delivered the men and the goods were shot down shortly after the drops, but at least their valuable cargo had reached the people who needed it. So it was at first thought in England. Actually, in late 1941 and early 1942, counterintelligence units of the German *Abwehr* stationed in Holland succeeded by D/F in locating a series of illegal radio transmitters of the Dutch underground and in capturing some of the operators. The Germans gradually substituted their own operators by blandly informing London that the old operator was not in good shape and the "underground" had supplied some new ones. This was counterintelligence at its wiliest. Playing the part of the Dutch underground on the air, the Nazis sucked into their maw many of the valiant volunteers and much of the equipment which was intended for their own destruction, thus effectively neutralizing part of the underground effort. This also accounted for the bombers being shot down after and not before they had delivered their supplies. Nazi control of Northpole was finally ended when two of the captured agents succeeded in escaping and in reaching England.

German D/Fing, which was at all times excellent, must also in great measure be given the credit for the initial breakthrough which caused the downfall of the major Soviet networks in Europe during World War II. By mid-1941, radio interception stations of German counterintelligence had recorded and examined a sufficient number of enciphered messages emanating from what were obviously illegal transmitters in Western Europe to realize that an extensive Soviet network was pumping information out of the German-occupied territories. The German D/Fing was dogged, unremitting and systematic. The Soviets, it is true, made the job easier for the Germans by requiring their operators to transmit for very long periods of time, since the intelligence to be reported was vital and extensive.

Just how significant the D/Fing technique has been for counter-intelligence is clear when one realizes that in this case the Germans had not the slightest clue as to the identity or whereabouts of any of the many Soviet agents who were gathering information of such interest to Moscow that five or more transmitters were keeping the air waves hot with it. Nor could the Germans make the slightest progress in breaking the ciphers used in these messages. The only possible way in which they could hope to close in on this unseen and unknowable spy system was by physically locating the radio transmitters into which the information was being fed. It was also a case of pinpointing a location not merely within a city but within an area of many thousands of square miles.

In a period of a little less than a year, from the fall of 1941 until the summer of 1942, *Abwehr* direction-finding units managed to locate three of the most important Soviet illegal radio stations and to apprehend the personnel of all three (since they were usually taken by surprise while transmitting). Two of the stations were in Belgium and one in France. Once the operators began to talk, and many of them gave out the most vital information about their networks under "persuasion" on the part of the Germans, the latter were, of course, able to get on the track of the agents and informants whose information had kept the radios so busy. With the assistance of one of the operators arrested in Belgium, the Germans tracked down the Schulze-Boysen-Harnack group in Berlin, described in the previous chapter. As in the Northpole case, the Germans kept some of the Soviet radios active for a time and succeeded in fooling Moscow long enough to smoke out further collaborators with Moscow's unwitting assistance.

As a result of these losses, and because it was by then too dangerous, if not impossible, to establish new illegal radio transmitters in Germany or German-occupied territory, the Soviets concentrated from 1942 onward on making Switzerland their communications base. Since the Soviets had no diplomatic representation in Switzerland, it was again necessary to resort to illegal transmitters. Many of them were eventually located and closed down as a result of Swiss D/Fing.

This account by no means exhausts the whole gamut of human and technical measures which counterintelligence has at its disposal. Much of its basic work is accomplished in the unglamorous area of its files, which constitute the backbone of any counterintelligence effort. One of the greatest advances in the administration of counterintelligence work has been the partial mechanization of file systems, which facilitates the quick and accurate recovery of world-wide counterintelligence information.

While much of the daily work of counterintelligence is laborious and humdrum, its complex and subtle operations are very much like a gigantic chess games that uses the whole world for its board.

9

Volunteers

The piercing of secrets behind the Iron and Bamboo Curtains is made easier for the West because of the volunteers who come our way.

We don't always have to go to the target. Often it comes to us through people who are well acquainted with it. While this is not a one-way street, the West has gained far more in recent years from volunteers than its opponents have. A reason for this change is the growing discontent with the system inside the Soviet Union, the satellite nations and Communist China, and some relaxation of the controls of Stalin's day. People know more, and they want more and they travel more.

These volunteers are either refugees and defectors who cross over the frontiers to us or they are people who remain "in place" in order to serve us from within the Communist societies.

Information from refugees is often piecemeal and scattered, but for years it has added to our basic fund of knowledge, particularly about Soviet satellites in Europe. The Hungarian Revolution in 1956 sent over a quarter of a million refugees fleeing westward. They brought us up to date on every aspect of technical, scientific and military achievement in Hungary and gave us an excellent forecast of likely capabilities for years to come. Among the hundreds of thousands of refugees who have come over from East Germany, other satellites and Communist China since the end of World War II, many have performed a similar service.

The term "defector" is often used in the jargon of international relations and intelligence to describe the officials or highly knowledgeable
citizens, generally from the Communist bloc, who leave their country
and come to the West. It is, however, a term that is resented, and properly so, by persons who repudiate a society which they leave in order to
join a better one.

I do not claim that all so-called defectors have come to the West for
ideological reasons. Some come because they have failed in their jobs;
some because they fear a shake-up in the regime may mean a demotion
or worse; some are lured by the physical attractions of the West, human
or material. But there is a large band who have come over to us from
Communist officialdom for highly ideological reasons. They have been
revolted by life in the Communist world and yearn for something better.
Hence, for these cases I use the term "defector" sparingly and then with
apology. I prefer to call them "volunteers."

If the man who comes over to us belonged to the Soviet hierarchy, he
may well know the strengths and weaknesses of the regime, its factions,
its inefficiencies and its corruption. If a specialist, he would know its
achievements in his chosen field. Volunteers may be soldiers, diplomats,
scientists, engineers, ballet dancers, athletes and, not infrequently, intelligence officers. Behind the Iron Curtain there are many dissatisfied persons unknown to us who seriously consider flight. Some of them hesitate
to take the final step, not because they have qualms about forsaking an
unsatisfying way of life, but because they are afraid of the unknowns that
await them.

The answer to this is to make it clear that they are welcome and will
be safe and happy with us. Every time a newly arrived political refugee
goes on the air over the Voice of America and says he is glad to be here
and is being treated well, other officials behind the Iron Curtain who
were thinking of doing the same thing will take heart and go back to fig-
uring out just how they can get themselves appointed as trade representatives in Oslo or Paris. Short-term visitors to the West from the Soviet
bloc would probably volunteer in far greater numbers were it not for the
Soviet practice of often keeping wives and children behind as hostages.

Oleg Lenchevsky, the Soviet scientist who sought asylum in Britain
in May of 1961 while he was studying there on a UNESCO fellowship,
tried in vain to get Khrushchev to permit his wife and two daughters,
whom he had left behind in Moscow, to leave the country and join him.
His personal appeal, in the form of a letter to Khrushchev, was published

in many Western newspapers. Khrushchev, of course, did not relent. He couldn't because he well knew that if he ever let Lenchevsky's family out of Russia, it would only set off a wave of defectors with families, all in hopes of the same treatment.

One of Lenchevsky's reasons for defecting was unusual, but symptomatic enough. He claimed that after years of suppressing his religious feelings he had suddenly felt the need of church and had been relieved to be able to attend services in Britain. He did not mention this in his letter to Khrushchev, but what he did mention was his discovery while in England of the contents of the Universal Declaration of Human Rights adopted by the General Assembly of the United Nations in 1948. Although all the signatories to this declaration, the Soviets included, agreed to its publication in every civilized country of the world, it had never seen the light of day in Soviet Russia. "Surely," Lenchevsky wrote Khrushchev,

> now, thirteen years later, when the liberty, fraternity, equality and happiness of all people have been proclaimed as our ideals in the new program of the Communist party, it is high time to put into practice these elementary principles of interhuman relations that are contained in the Universal Declaration of Human Rights.

A frequent cause for unrest among scientists, artists and writers behind the Iron Curtain is quite naturally the lack of freedom of inquiry in their fields, the imposition of political theses on their work which even goes so far as to reject ideas that tend to conflict with Marxist views of the world. In some fields an honest Soviet scientist stands in about the same relation to the state as Galileo did to the Inquisition 350 years ago (recant or be punished). The Lysenko controversy was one of the most publicized affairs in which laboratory science and Marxist ideology clashed head on, and Marxism, of course, won. The theories of biologists who opposed Lysenko and genetic findings which emphasized the importance of heredity were rejected by a state which rules that man can be transformed by his environment. The outstanding Soviet chemist, Dr. Mikhail Klochko, a Stalin Prize winner, who defected in Canada in 1961, wrote:

> The Soviet Encyclopedia had appeared with an article on physical chemistry written by scientists senior to me, which was both biased and ludicrous. At a meeting I pointed this out. Many persons told me later that although they agreed with me, they thought I should not

get into trouble with these powerful men. But this event merely reinforced the conviction I now had that I must leave the Soviet Union if
ever I was to achieve my full potentialities as a scientist.[1]

I believe that, given a free opportunity to leave, the number of people
who today would move out from behind the Iron and Bamboo Curtains
would be, without exaggeration, astronomical. The total from the end
of World War II until the end of 1961, the year the Berlin Wall went up,
was over 11 million, and most of them had not been given the opportunity to leave; they took it. The available figures, which include war-
displaced persons who did not wish to return to their homelands behind
the Curtain after the war was over, as well as refugees and defectors, are
by area of origin, estimated as follows:

East Germany	3,600,000
Baltic states	200,000
European satellites	1,783,000
Communist China	3,000,000
Asian satellites	2,000,000
Soviet Russia	1,000,000
Total	11,583,000

The Communists will go to great lengths to prevent the defection of any
person whom they regard as "valuable" to them or of possible use to us.
Western scientists at international conferences attended by Soviet and
satellite delegations have frequently tried to start friendly conversations
with one or another of the members of such delegations decked out as
chemists or meteorologists, only to stumble upon the one man who does
not know the first word about the subject in which the delegation was
supposed to be expert. He is the KGB security man who has been sent
along solely for the sake of keeping an eye on the bona fide scientists in
the delegation, to see that they don't talk out of turn and, above all, that
they don't make a break for freedom.

The Chinese Communists carefully limit the amount of fuel in the
tanks of their military planes before the latter go on training missions or
maneuvers so that a pilot who might take it into his head while aloft to
steer for Formosa and freedom cannot reach his goal. Even so, a few

[1] *This Week* Magazine, December 31, 1961.

President Kennedy and Mr. Dulles at the inauguration of the new CIA Headquarters in November, 1961.

Below, an aerial photo of the Headquarters in Virginia.

WIDE WORLD

Benjamin Franklin dictating to Edward Bancroft, his secretary-assistant who was an espionage agent for Britain during the Revolution. CULVER PICTURES, INC.

Major Allan Pinkerton (left), who organized an espionage system for the U.S. early in the Civil War, with President Abraham Lincoln and Maj. Gen. J. A. McClernand. CULVER PICTURES, INC.

Henry L. Stimson, when Secretary of State in 1929, closed down the so-called Black Chamber. CULVER PICTURES, INC.

Charles Evans Hughes, then Secretary of State, with delegates to the 1921 Disarmament Conference. At this time American cryptographers had broken the Japanese diplomatic code. CULVER PICTURES, INC.

Richard Sorge, German newspaperman in Tokyo who ran a spy ring for Soviet Russia in Japan in the early days of World War II. WIDE WORLD

Klaus Fuchs, who gave atomic secrets to Soviet Russia, arriving in Easy Germany after release from a British prison. WIDE WORLD

The Soviet agent, Frank Jackson, who murdered Leon Trotsky in Mexico City in 1940. WIDE WORLD

David Greenglass, member of the
Rosenberg atomic spy ring, after his
arraignment in 1950. WIDE WORLD

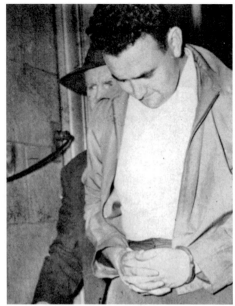

Rudolf Abel, Soviet spy who
masqueraded as a photographer
in Brooklyn. WIDE WORLD

U-2 Pilot Francis Gary Powers
before the Senate Armed Services
Committee. WIDE WORLD

This picture of the San Diego, California, naval air station, taken at 70,000 feet, illustrates the scope of U-2 aerial surveillance at great heights. WIDE WORLD

Henry Cabot Lodge, U.S. Ambassador to the UN, demonstrating to the Security Council in May, 1960, how a microphone had been concealed in a wooden carving of the Great Seal of the U.S. in the American Embassy in Moscow. WIDE WORLD

John Vassall, former British Admiralty clerk, convicted in 1962 of spying for Russia. WIDE WORLD

Oleg Penkovsky in military court in Moscow where he was sentenced to death for espionage in May, 1963. WIDE WORLD

years ago one of their pilots happened to make it. The first night after he landed he was put up at a farm out in the country. The next morning he was asked how he had slept during his first night of freedom. He hadn't slept well, he said, because of the noise. "Noise?" he was asked. "Out here in the country? What noise?" It turned out that the clucking of the chickens had kept him awake. He wasn't used to it. Barnyard noises apparently are on the wane on the mainland.

On the other hand, the fate of some who have gone from our side over to the Soviets would not serve as a particularly good advertisement for further defections in that direction. Some of them recently have talked to Western visitors and have admitted, without prompting, that their lot is an unhappy one and that they have no future. The scientific defectors, like the atomic physicist Pontecorvo, who continue to be useful to the Soviets in their technological efforts, seem to fare better than the others, and sometimes even receive high honors, as Pontecorvo did when he was awarded the Lenin Prize. The Burgesses and MacLeans, the Martins and Mitchells, had their day of publicity and then eked out a dull living, some as "propaganda advisers."

Often "defectors" from the Communist side are not exactly what they seem. Some, for example, have been working as agents "in place" behind the Curtain for long periods of time before defecting and only come out because they or we feel that the dangers of remaining inside have become too great.

People who volunteer "in place" have many ways of doing so, even though the isolation, the physical barriers and the internal controls of the Soviet bloc are all supposed to prevent this kind of thing from happening. It is possible, also, for them to communicate safely with the West in a number of ways—surprisingly enough, even by mail, as long as the address of the recipient looks harmless and the identity of the sender within the bloc remains concealed. Soviet bloc censorship cannot possibly inspect every piece of mail passing to and fro over their borders—the volume is too great. Even if a letter is censored or intercepted, it need give no clue whatever about the identity of the sender if proper security precautions are followed. Various radio stations in Western Europe that broadcast to the Soviet bloc solicit comments and fan mail from listeners and usually supply a postbox to which such mail can be sent. They receive many letters from behind the Iron Curtain. If a volunteer who has mailed out information succeeds later in reaching the West, he then, of course, finds a ready welcome there.

Some very helpful and important defectors have been diplomats or intelligence officers under diplomatic cover. It is, of course, a relatively simple matter for them while posted abroad in a free country to walk out of their jobs one fine day and go to the Foreign Office of the country to which they are accredited or a Western embassy and request protection. In the West, whenever this happens and when the motives of the defecting diplomat appear to be bona fide, the requested protection and material assistance needed until the diplomat can find a new livelihood in his new home are usually granted.

If there is any hesitancy in extending these privileges, it is because the Soviets have from time to time mounted phony defections, which is rather an unsatisfactory way of planting an agent but may have incidental benefits. The phony "defector," when interviewed by persons in the country to which he has "defected," may pick up and be able to send back a certain amount of information, especially concerning what is known or not known about his own country. A further and final step in such phony defections is that the defector may eventually "redefect." One day he will announce that he is disillusioned with the West, that it is not as represented, he repents of his sins and wants to go home even if he is to be punished for his original defection. This provides some propaganda repercussion, is embarrassing to the country of haven, and is a convenient way for the defector, who was really an agent, to return home and report on the information he has been assembling. But this is the exception, and the Soviets have not tried it much lately, chiefly, I think, because it has not worked well. It has usually been possible to discover quite early in the day whether the man was bona fide or not. In some cases, phony defectors have confessed that they were planted.

Soviet and satellite intelligence officers, like the diplomats, also have the advantage of posts and of trips abroad, and some use such occasions to make the break they may long have been contemplating. Their defections are regarded as most serious losses by the Soviets. They may go to great lengths to prevent such defections from happening, even to using violence to force the return of a potential defector, not to mention reprisals of various kinds should the defection succeed or the defector's family remain under Soviet control.

The reader may recall the sensational news photos in 1954 which showed a Soviet goon squad strong-arming the wife of defector Vladimir Petrov, KGB Chief in Australia, in an attempt to get her on a plane and take her back to Russia against her will. Only the quick intervention of the Australian police saved Mrs. Petrov from being abducted.

For these reasons the defection of intelligence officers is often carried out with much less fanfare than those of more public personages like diplomats or scientists. The Soviet or satellite intelligence officer also usually has the advantage of knowing in some degree how to get in touch with his "opposite numbers" in the West. After all, part of his job was to probe for such information. When he picks up and leaves, it is likely that he will head for a Western intelligence installation rather than for a diplomatic establishment or the nearest police station because he can be fairly certain of his welcome there and that his defection will be handled most securely.

The defection of a staff intelligence officer of the opposition is naturally a break for Western counterintelligence. It is often the equivalent, in the information it provides, of a direct penetration of hostile headquarters for a period of time. One such intelligence "volunteer" can literally paralyze the service he left behind for months to come. He can describe the internal and external organization of his service and the work and character of many of his colleagues at headquarters. He can identify intelligence personnel stationed abroad under cover. Best of all, he can deliver information about operations. Yet he may not know the true identity of a large number of agents for the reason that all intelligence services compartmentalize such information. No one knows true identities except the few officers intimately concerned with a case.

The West has been singularly fortunate in having many such defectors come over to its side in the course of recent history. In 1937 two of Stalin's top intelligence officers stationed abroad defected rather than return to Russia to be swallowed up in the purge of the NKVD, which followed the purges of the party and of the Army. One was Walter Krivitsky, who had been chief of Soviet intelligence in Holland. He was found dead in a Washington hotel in 1941, shot presumably by agents of the Soviets who were never apprehended. The story that he committed suicide seems most unlikely. The second was Alexander Orlov, who had been one of the NKVD chiefs in Spain at the time of the Civil War. Unlike Krivitsky, he has managed to elude Soviet vengeance and has published a number of books, one on Stalin's crimes and another on Soviet intelligence.

An early postwar Soviet defector was Igor Gouzenko, whom I mentioned earlier. Gouzenko was a military intelligence officer in charge of codes and ciphers in the Soviet Embassy in Ottawa. Thanks in some measure to clues he brought with him, part of the international atomic spy ring which the Soviets had been running during and after the last years of the war was uncovered.

Following the liquidation of Beria shortly after Stalin's death in 1953, it was clear to officers of the Soviet Security Service that anyone who had served under him was in jeopardy. The new regime would not feel sure of the loyalty of old-timers who knew too much. The new regime could also make itself more popular by going through the motions of wiping out the hated secret police of a previous regime and quietly putting its own loyal adherents in their places.

Among the major defectors to the West at that time were Vladimir Petrov, whom I have just mentioned; Juri Rastvorov, an intelligence officer stationed at the Soviet mission in Japan; and Peter Deriabin, who defected from his post in Vienna. All these men had at some time been stationed at intelligence headquarters in Moscow and possessed valuable information that went far beyond their assignments at the time they defected. Deriabin later told his story in a book called *The Secret World*.

In recent years, two defections of a special kind have involved Soviet intelligence personnel employed on assassination missions. Nikolay Khokhlov was sent from Moscow to West Germany in early 1954 to arrange for the murder of a prominent anti-Soviet émigré leader, Georgi Okolovich. Khokhlov told Okolovich of his mission and then defected. At Munich in 1957, Soviet agents tried without success to poison Khokhlov. In the fall of 1961, Bogdan Stashinski defected in West Germany and confessed that on Soviet orders he had murdered the two Ukrainian exile leaders Rebet and Bandera some years earlier in Munich.

In 1959, Soviet diplomat Aleksandr Kaznacheev defected in Burma, where he had been stationed in the embassy. While Kaznacheev was not a staff member of Soviet intelligence, he was a "coopted worker" and was used in intelligence work whenever his position as a diplomat enabled him to perform certain tasks with less risk of discovery than his colleagues in the intelligence branch. His candid book describing what went on in the Soviet embassy in Rangoon[2] has done a

---

[2] *Inside a Soviet Embassy*, J. B. Lippincott Co., 1962.

great deal to debunk the picture of Soviet skill and American incompetence previously impressed on the American public in the book *The Ugly American*.

The latest and one of the most advertised defections of a Soviet intelligence officer took place in early February of 1964, when an "expert" attached to the Soviet delegation to the Geneva Disarmament Conference, Yuri I. Nossenko, disappeared from view and was reported some days later by our own State Department to have requested asylum in the United States. Nossenko was a high-ranking staff officer of the KGB, presumably well-versed in security as well as in scientific matters. It was somewhat amusing in this case that the Soviets went to the Swiss police, before the official U.S. announcement was made, to ask for help in locating their missing man. They would hardly have done that in Stalin's day. It was tantamount to their saying: please help us keep our personnel under control, since we can't do it ourselves.

All the important intelligence "volunteers" have not been Soviets. Numerous high-ranking staff officers have defected from the satellite countries and were able to contribute information not only about their own services but about Soviet intelligence as well. Whatever impression of independence European satellite governments may try to give, they are, in matters of espionage, satrapies of the U.S.S.R. When agents of the satellite services come over to the West, they are a window on the policies and plans of the Kremlin.

Joseph Swiatlo, who defected in Berlin in 1954, had been chief of the department of the Polish intelligence service which kept tabs on members of the Polish Government and the Polish Communist party. Needless to say, he knew all the scandal about the latter, and the Soviets had frequently consulted with him.

Pawel Monat had been Polish military attaché in Washington from 1955 to 1958, after which he had returned to Warsaw and was put in charge of world-wide collection of information by Polish military attachés. He served in this job for two years before defecting in 1959. We will hear more of him later on.

Frantisek Tisler defected in Washington after having served as Czech military attaché there from 1955 to 1959. The Hungarian secret police officer, Bela Lapusnyik, made a daring escape to freedom over the Austro-Hungarian border in May, 1962, and reached Vienna in safety, only to die of poisoning apparently at the hands of Soviet or Hungarian agents, before he could tell his full story to Western authorities.

The Chinese defector, Chao Fu, who had been serving as the "security officer" in the Red Chinese embassy in Stockholm until he "disappeared" in 1962, was one of the first openly publicized cases of a defection from the Chinese Communist State Security Service. There are others.

What has brought these men and others over to our side is naturally a matter of great interest, not only to Western intelligence, but to any serious student of the Soviet system and of Soviet life. Gouzenko, for example, has told how he was gradually overcome by shame and repugnance as he began to realize that the U.S.S.R., while a wartime ally of Britain, Canada and the United States, was mounting a massive espionage effort to steal scientific secrets. This moral revulsion eventually led to his defection.

The postwar defectors were not in a similar situation because the Soviets after 1946 were no longer even pretending to be our friends. Every Soviet official was well indoctrinated on this point and could not easily survive in his job if he had any soft feelings about the "imperialists." Nevertheless, feelings akin to those which stirred Gouzenko seem to have moved others. Most defectors have suffered some kind of disillusionment or disappointment with their own system.

When one studies the role the intelligence services play in the Soviet world and their closeness to the centers of power, it is not surprising that the Soviet intelligence officer gets an inside look, available to few, of the sinister methods of operation behind the façade of "socialist legality." To the intelligent and dedicated Communist, such knowledge comes as a shock. One defector has told us, for example, that he could trace the disillusionment which later led to his own defection back to the day when he found out that Stalin and the NKVD, and not the Germans, had been responsible for the Katyn massacre (the murder of about ten thousand Polish officers during World War II). The Soviet public still does not know the truth about this or most of the other crimes of Stalin. But once a man is aware of realities, "loss of faith" in the system within which he is working, coupled often with personal disappointments, seems to be the powerful driving factor in defections.

The names mentioned here by no means exhaust the list of all those who have left the Soviet intelligence service and other Soviet posts. Some of the most important and also some of the most recent defectors have so far chosen not to be "surfaced," and for their own protection must remain unknown to the public. They are making a continual contribution to the inside knowledge of the work of the Soviet intelligence and security apparatus and to exposing the way in which the subversive war is being carried on against us by Communism.

The United States in particular has always been a haven for those seeking to leave tyranny and espouse freedom. It will always have a welcome for those who do not wish to continue to work for the Kremlin.

10

Confusing the Adversary

In intelligence, the term "deception" covers a wide variety of maneuvers by which a state attempts to mislead another state, generally a potential or actual enemy, as to its own capabilities and intentions. Its best-known use is in wartime or just prior to the outbreak of war, when its main purpose is to draw enemy defenses away from a planned point of attack, or to give the impression that there will be no attack at all or simply to confuse the opponent about one's plans and purposes.

As a technique, deception is as old as history. Notable instances come down to us from Homer and Thucydides: the Trojan horse that led to the fall of Troy and the strategy of the Greeks attacking Syracuse in 415 B.C. In the latter case the Greeks infiltrated a plausible agent into the ranks of the Syracusans, lured them to attack the Greek camp at some distance from the city and meanwhile put their whole army on board ship and sailed for Syracuse, which was left practically undefended.

During the kind of peace we now call Cold War, various other forms of deception, including political deception, are being practiced against us by the Soviets, often involving the use of forgeries. Deception took an even less subtle form in Cuba when the Soviets, while vigorously denying any complicity in installing their intermediate-range or offensive-type missiles, were caught in the act.

As a strategic maneuver, deception generally requires lengthy and careful preparation. Intelligence must first ascertain what the enemy thinks and what he expects, because the misleading information which is going to be put into his hands must be plausible and not outside the practical range of plans that the enemy knows are capable of being put into operation. Intelligence must then devise a way of getting the deception to the enemy. Success depends on close coordination between the military command and the intelligence service.

After the Allies had driven the Germans out of North Africa in 1943, it was clear to all that their next move would be into southern Europe. The question was where. Since Sicily was an obvious steppingstone and was in fact the Allied objective, it was felt that every effort should be made to give the Germans and Italians the impression that the Allies were going to by-pass it. To have tried to persuade the Germans that there was to be no attack at all or that it was going to move across Spain was out of the question, for these maneuvers would not have been credible. The deception had to point to something within the expected range.

For quick and effective placement of plausible deception directly into the hands of the enemy's high command, few methods beat the "accident," so long as it seems logical and has all the appearances of being a wonderfully lucky break for the enemy. Such an accident was cleverly staged by the British in 1943 before the invasion of Sicily, and it was accepted by the Germans at the time as completely genuine. Early in May of that year the corpse of a British major was found washed up on the southwest coast of Spain near the town of Huelva, between the Portuguese border and Gibraltar. A courier briefcase was still strapped to his wrist containing copies of correspondence to General Alexander in Tunisia from the Imperial General Staff. These papers clearly hinted at an Allied plan to invade southern Europe via Sardinia and Greece. As we learned after the war, the Germans fully believed these hints. Hitler sent an armored division to Greece, and the Italian garrison on Sicily was not reinforced.

This was perhaps one of the best cases of deception utilizing a single move in recent intelligence history. It was called "Operation Mincemeat," and the story of its execution has been fully told by one of the main planners of the affair, Ewen Montagu.[1] It was a highly sophisticated feat, made possible by the circumstances of modern warfare and the techniques if modern science. There was nothing illogical about the possibility that a plane on which an officer carrying important documents was a

[1] *The Man Who Never Was* (Philadelphia: J.B. Lippincott Co., 1954).

passenger could have come down, or that a body from the crash could have been washed up on the Spanish shore.

Actually, the body of a recently dead civilian was used for this operation. He was dressed in the uniform of a British major; in his pockets were all the identification papers, calling cards and odds and ends necessary to authenticate him as Major Martin. He was floated into Spain from a British submarine, which surfaced close enough to the Spanish coast to make sure that he would reach his target without fail. And he did.

"Overlord," the combined Allied invasion of Normandy, in June, 1944, also made effective use of deception—in this case not an isolated ruse but a variety of misleading maneuvers closely coordinated with each other. They succeeded, as is well known, in keeping the Germans guessing as to the exact area of the intended Allied landing. False rumors were circulated among our own troops on the theory that German agents in England would pick them up and report them. Radio channels to agents in the French underground were utilized to pass deceptive orders and requests for action to back up the coming Allied landings; it was known that certain of these agents were under the control of the Germans and would pass on to them messages received from the Allies. Such agents therefore constituted a direct channel to the German intelligence service. In order to make the Germans think that the landings would take place in the Le Havre area, agents in the vicinity were asked to make certain observations, thereby indicating to the Germans a heightened Allied interest in fortifications, rail traffic, etc. Lastly, military reconnaissance itself was organized in such a way as to emphasize an urgent interest in places where the attack would not come. Fewer aerial reconnaissance sorties were flown over the Normandy beaches than over Le Havre and other likely areas. Rumors were spread of a diversionary attack on Norway to prevent a concentration of forces in the North of France.

There are essentially two ways of planting deceptive information with the enemy. One can stage the kind of accident the British did in Spain. Such accidents are plausible because they do, after all, frequently occur solely as a result of the misfortunes of war. History is full of instances where couriers loaded with important dispatches fell into enemy hands. The other way is to plant an agent with the enemy who is ostensibly reporting to him about your plans as the Athenians did at Syracuse. He can be a "deserter" or some kind of "neutral." The problem, as in all counterespionage penetrations, is to get the enemy to trust the agent. He cannot simply turn up with dramatic military information and expect to be believed unless he can explain his motives and how he got his information.

A wholly modern deception channel came into being with the use of radio. For example, a parachutist lands in enemy territory equipped with a portable transmitter and is captured. He confesses he has been sent on a mission to spy on enemy troop movements and to communicate with his intelligence headquarters by radio. Such an agent stands a good chance of being shot after making this confession; he may be shot before he has a chance to make it. The probability is high, however, that his captors will decide he is more useful alive than dead because his radio provides a direct channel for feeding deception to the opponent's intelligence service. If the intelligence service which sent the agent knows, however, that he has been captured and is under enemy control, it can continue to send him questions with the intent of deceiving the other side. If it asks for a report on troop concentrations in sector A, it gives the impression that some military action is planned there. This was one tactic used by the Allies in preparation for the Normandy landings.

A lesser and essentially defensive kind of deception involves the camouflaging of important targets. To deceive Nazi bombers during World War II, airfields in Britain were made to look like farms from the air. Sod was placed over the hangars and maintenance shacks were given the appearance of barns, sheds and outbuildings. Even more important, mock-ups were set up in other areas to look like real airfields with planes on them. Elsewhere mocked-up naval vessels were stationed where the real might well have been.

The mounting of strategic deception calls for the close cooperation and high security of all parts of government engaged in the effort. For a democratic government this is difficult except under wartime controls.

For the Soviets, of course, the situation is somewhat easier. With their centralized organization and complete control of the press and of dissemination of information within their country or to foreign countries from the U.S.S.R., they can support a deception operation far more efficiently than we can. Often the Soviets put armaments on display with a certain amount of fanfare in order to draw attention away from other armaments they may have in their arsenal or may plan to have. Sometimes they exhibit mock-ups of planes and other equipment, which may never see the light of day as operational types.

For example, on Aviation Day in July, 1955, in the presence of diplomatic and military representatives in Moscow there was a "fly-by" of a new type of Soviet heavy bomber. The number far exceeded what was thought to be available. The impression was thus given that many more had lately come off the assembly line and that the Soviets were therefore committed to an increasing force of heavy bombers. Later it was surmised that the same squadron had been flying around in circles, reappearing every few minutes. The purpose was to emphasize Soviet bomber production. In fact, they were soon to shift the emphasis to missiles.

Deception can also use social channels. A Soviet diplomat drops a remark in deepest confidence to a colleague from a neutral country at a dinner party, knowing that the neutral colleague also goes to British and American dinner parties. This "casual remark" was contained in a directive from the Soviet Foreign Office. When it is studied in intelligence headquarters somewhere in the West, it is found to agree in substance with something said by a Soviet official at a cocktail party ten thousand miles away. Thus, the two remarks seem to confirm each other. In reality both men were speaking as mouthpieces in a program of political deception which the Soviets coordinate with their ever-shifting plots in Berlin, Laos, the Congo, Cuba and whatever is next on the program.

One of the most successful long-range political deceptions of the Communists convinced gullible people in the West before and during World War II that the Chinese people's movement was not Communistic, but a social and "agrarian" reform movement. This fiction was planted through Communist-influenced journalists in the Far East and penetrated organizations in the West.

The Soviets have centralized the responsibility for planning and launching deception operations in a special department of the State Security Service (KGB) known as the "Disinformation Bureau." In recent years this office has been particularly busy formulating and distributing what purport to be official documents of the United States, Britain and other countries of the Free World. Its intention is to misstate and misrepresent the policies and purposes of these countries. In June of 1961, Mr. Richard Helms, a high official of the Central Intelligence Agency, presented the evidence of this activity to a Congressional committee. Out of the mass of forgeries available, he selected thirty-two particularly succulent ones, which were fabricated in the period 1957–60.

He pointed out that the Russian secret service has a long history of forging documents, having concocted the *Protocols of Zion* over sixty years ago to promote anti-Semitism. The Soviets have been adept pupils of their czarist predecessors. Their forgeries nowadays, he pointed out, are intended to discredit the West, and the United States in particular, in the eyes of the rest of the world; to sow suspicion and discord among the Western allies; and to drive a wedge between the peoples of non-Communist countries and their governments by promoting the notion that these governments are the puppets of the United States.

The falsified documents include various communications purporting to be from high officials to the President of the United States, letters to and from the Secretary of State or high State Department, Defense Department and USIA officials. To the initiated, these documents are patent fabrications; while some of the texts are cleverly conceived, there are always a great number of technical errors and inconsistencies. Unfortunately, these are not apparent to the audiences for which the letters are intended, generally the peoples of the newly independent nations. The documents are prepared for mass consumption rather than the elite. One of the most subtle, supposedly part of a British Cabinet paper, wholly misrepresented the U.S. and British attitude with respect to trade-union policies in Africa.

A typical Soviet forgery which appeared in an English-language newspaper in India consisted of two spurious telegrams allegedly sent by the American Ambassador in Taipeh to the Secretary of State in Washington commenting on various wholly fictitious proposals for doing away with Chiang Kai-shek. In order to explain how the "telegrams" had fallen into their hands, the Soviets cleverly exploited the fact that a mob had shortly before raided our embassy in Taipeh.

The forgery technique is particularly useful to the Communists because they possess the means for wide and fast distribution. Newspapers and news outlets are available to them on a world-wide basis. While many of these outlets are tarnished and suspect because of Communist affiliations, they are nevertheless capable of placing a fabrication before millions of people in a short time. The denials and the pinpointing of the evidence of fabrication ride so far behind the initial publication that the forgeries have already made their impact in spreading deception. On the other hand, the technique of forgery is not so readily available to Western intelligence in peacetime, for, quite apart from ethical considerations, there is too much danger of deceiving and misleading our own people and our free press.

When one deliberately misleads, sometimes friend as well as foe is misled. And later the deceiver may not be believed when he wishes to be. This is the situation of the Soviets today after Cuba.

Often the very fear of deception has blinded an opponent to the real value of the information which accidents or intelligence operations have placed in his hands.

As Sir Walter Scott wrote:

Oh, what a tangled web we weave,
When first we practice to deceive!

If you suspect an enemy of constant trickery, then almost anything that happens can be taken as one of his tricks. A collateral effect of deception, once a single piece of deception has succeeded in its purpose, is to upset and confuse the opponent's judgment and evaluation of other intelligence he may receive. He will be suspicious and distrustful. He will not want to be caught off guard.

On January 10, 1940, during the first year of World War II, a German courier plane flying between two points in Germany lost its way in the clouds, ran out of fuel and made a forced landing in what turned out to be Belgium. On board were the complete plans of the German invasion of France through Belgium, for which Hitler had already given marching orders. When the *Luftwaffe* major who had been piloting the plane realized where he had landed, he quickly built a fire out of brush and tried to burn all the papers he had on board, but Belgian authorities reached him before he could finish the job and retrieved half-burned and unburned documents to be able to piece together the German plan.

Some of the high British and French officials who studied the material felt that the whole thing was a German deception operation. How could the Germans be so sloppy as to allow a small plane to go aloft so close to the Belgian border in bad weather with a completely detailed invasion plan on board? This reasoning focused on the circumstances, not on the contents of the papers. Churchill writes that he opposed this interpretation. Putting himself in the place of the German leaders, he asked himself what possible advantage there was at that moment in perpetrating a deception of this sort, i.e., alerting Belgium and Holland by faking invasion plans. Obviously, none. As we learned after the war, the invasion of Belgium, which had been set for the sixteenth of January—six days after the plane came down—was postponed by Hitler primarily because the plans had fallen into the Allies' hands.

Accidents like this are not the only events that raise the specter of deception. It has already been pointed out that if you send a deception agent to the enemy, you have to make him credible. Bona fide windfalls have sometimes been doubted and neglected because they were suspected of being deception. This happened to the Nazis late in World War II in the case of "Cicero," the Albanian valet of the British Ambassador to Turkey. He had succeeded in cracking the ambassador's private safe and had access to top-secret British documents on the conduct of the war. One day he offered to sell them to the Germans as well as to continue supplying similar documents.

His offer was accepted, but some of Hitler's experts in Berlin could never quite believe that this wasn't a British trick. Their reasons, however, were more complex than in the cases where deception alone is feared. The incident is also an excellent example of how prejudice and preconception can cause failure properly to evaluate valid intelligence. For one thing, the Cicero documents gave evidence of the massive Allied offensives to come and the growing power of the Allies—information which collided head on with illusions cherished in the highest Nazi circles. Second, competition and discord among different organs of the German government prevented it from making a sober analysis of this source. The intelligence service under Himmler and Kaltenbrunner and the diplomatic service under Ribbentrop were at odds and, as a result, if Kaltenbrunner thought information was good, Ribbentrop automatically tended to think it was bad. An objective analysis of the operational data was out of the question in a situation where rival cutthroats were vying for position and prestige. In the Cicero case, Ribbentrop and the diplomatic service suspected deception. The net effect was that, as far as can be ascertained, the Cicero material never had any appreciable influence on Nazi strategy. Contrary to the general impression, there is also no evidence that the Nazis gained from Cicero any information about the planned invasion of Europe except possibly the code work for the operation—"Overlord."

A further ironical twist to this famous case is that the Nazi intelligence service paid this most valuable agent hundreds of thousands of pounds in counterfeit English notes. Cicero has been trying ever since to get restitution from the German government for services rendered—in real money.

11

How Intelligence Is Put to Use

Information gathered by intelligence services or compiled by the analyst is of little use unless it is got into the hands of the "consumers," the policymakers. This must be done promptly and in clear, intelligible form so that the particular intelligence can easily by related to the policy problem with which the consumers are then concerned.

These criteria are not easily met, for the sum total of intelligence available is very great on many subjects. Thousands of items come into CIA headquarters every day, directly or through other agencies of government, particularly the State and Defense departments. Many other items are added from the research work of scholars. When we consider all we need to know about happenings behind the Iron Curtain and in over a hundred other countries, this volume is not surprising. Anywhere in the world events could occur which might affect the security of the United States. How is this mass of information handled by the various collection agencies, and how is it processed in the State Department, the Defense Department and the CIA?

Between these three agencies there is immediate and often automatic exchange of important intelligence data. Of course, someone has to decide what "important" means and determine priorities. The sender of an intelligence report (who may be any one of our many officials abroad—diplomatic, military or intelligence) will often label it as being of a certain importance, but the question of priority is generally decided on the receiving end. If a report is of a particularly critical character, touching on the danger of hostilities or some major threat to our national security, the sender will place his message in channels that provide for automatic dissemination to the intelligence officers in the State and Defense departments and the CIA. The latter, as coordinator of foreign intelligence, has the right of access to all intelligence that comes to any department of our government. This is provided for by law.

There is a round-the-clock watch for important intelligence coming into the State and Defense departments and the CIA. During office hours (which in intelligence work are never normal), designated officers scan the incoming information for anything of a critical character. Through the long night hours, special watch officers in the three agencies do the monitoring. They are in close touch with each other, come to know each other well and continually exchange ideas about the sorting of clues to any developing crisis. In the event that any dramatic item should appear in the incoming nightly stream of reports, arrangements have been made as to the notification of their immediate chiefs. The latter decide who among the high policy officials of government—from the President at the top to the responsible senior officers in State, Defense and the CIA—should be alerted. The watch officers also follow the press service and radio reports, including those of Soviet and Chinese Communist origin. News of a dramatic, yet open, character—the death of a Stalin, a revolt in Iraq, the overthrow of a political leader—may first become known through public means of communication. Our officials abroad today have access to the most speedy means of transmission of reports from our embassies and our overseas installations, but these messages must go through the process of being enciphered and deciphered. As a result, news flashes sometimes get through first.

After there has been an important incident affecting our security, one that has called for policy decisions and actions, there is usually an intelligence postmortem to examine how effectively the available information was handled and how much forewarning had been given by intelligence. Incidents such as the Iraqi revolution of 1958 or the erecting of the wall dividing Berlin on August 13, 1961, required such treatment, since neither had been clearly predicted through intelligence channels. The purpose of the postmortem is to obtain something in the nature of a batting average of the alertness of intelligence services. If there has been a failure, either in prior warning or in handling the intelligence already at hand, the causes are sought and every effort is made to find means of improving future performance.

The processing of incoming intelligence falls into three general categories. The first is the daily and hourly handling of current intelligence. The second is the researching of all available intelligence on a series of subjects of broad interest to our policymakers; this might be given the name "basic intelligence." For example, one group of analysts may work on the information available on the Soviet economy, another with its agriculture, a third with its steel and capital goods production and still another with its aircraft and missile development. The third type of processing involves the preparation of an intelligence estimate, which is described below.

There is, of course, not time to submit every important item of current intelligence to detailed analysis before it is distributed to the policymakers. But "raw" intelligence is a dangerous thing unless it is understood for what it generally is—an unevaluated report, frequently sent off without the originator of the message being able to determine finally its accuracy and reliability. Hence the policymakers who receive such intelligence in the form of periodic bulletins (or as an isolated message if its importance and urgency require special treatment) are warned against acting on raw intelligence alone.

Bulletins, both daily and weekly, summarize on a world-wide basis the important new developments over the preceding hours or days; they include such appraisal as the sender may give or as the CIA is able to add in consultation with representatives of the other government intelligence agencies. These representatives meet frequently for that purpose, going over the items to be included in the daily bulletin. New information may still be added to the daily bulletin up until the early morning hours of the day on which it is issued. When this intelligence is sent forward, explanatory material is often included as to source, manner of acquisition and reliability. Some messages carry their own credentials as to authenticity; most do not.

In addition to the current raw intelligence reports and the "basic intelligence" studies, there are the position papers, generally called "national estimates." These are prepared by the intelligence community on the basis of all the intelligence available on a certain subject along with an interpretation of the "imponderables." Here we come to a most vital function of the entire work of intelligence—how to deal with the mass of information about future developments so as to make it useful to our policymakers and planners as they examine the critical problems of today and tomorrow. Berlin, Cuba, Laos; Communist aims and objectives; the Soviet military and nuclear programs; the economics of the U.S.S.R. and Communist China—the list could be almost indefinitely extended and is, of course, not exclusively concerned with Communist bloc matters. Sometimes estimates must be made on a crash basis. Sometimes, particularly where long-range estimates are involved, they are made after weeks of study.

One of the major reasons for the organization of the CIA was to provide a mechanism for coordinating the work of producing intelligence estimates so that the President, the Secretary of State and the Secretary of Defense could have before them a single reasoned analysis of the factors involved in situations affecting our national security. President Truman, who, in 1947, submitted the legislation proposing its creation, expressed in his memoirs the need for such a mechanism:

> The war taught us this lesson—that we had to collect intelligence in a
> manner that would make the information available where it was
> needed and when it was wanted, in an intelligent and understandable
> form. If it is not intelligent and understandable, it is useless.

He also describes the system by which intelligence was coordinated and passed on to policymakers:

> Each time the National Security Council is about to consider a cer-
> tain policy—let us say a policy having to do with Southeast Asia—it
> immediately calls upon the CIA to present an estimate of the effects
> such a policy is likely to have. The Director of the CIA sits with the
> staff of the National Security Council and continually informs as
> they go along. The estimates he submits represent the judgment of
> the CIA and a cross section of the judgments of all the advisory
> councils of the CIA. These are G-2, A-2, the ONI, the State Depart-
> ment, the FBI, and the Director of Intelligence of the AEC. The Sec-
> retary of State then makes the final recommendation of policy, and
> the President makes the final decision.[1]

[1] *Memoirs of Harry S. Truman* (New York: Doubleday & Co., Inc., 1958).

What President Truman refers to as "the advisory councils of the CIA" was established in 1950 as the Intelligence Advisory Committee, which later became the United States Intelligence Board (USIB) and is often referred to as "the intelligence community." USIB now has an additional member to those listed above—the head of the newly created Defense Intelligence Agency, which coordinates the work of Army, Navy and Air Force intelligence and is playing an increasingly important role in the intelligence community. So too is the intelligence unit of the State Department, whose head ranks as an Assistant Secretary of State. The USIB meets regularly every week and more frequently during crises or whenever any vital new item of intelligence is received. The Director of Central Intelligence, who is chairman of the board, is responsible for the estimates produced by the board. However, if any member dissents and desires his dissent to be recorded, a statement of his views is included as a footnote to the estimate that is finally presented to the President and interested members of the National Security Council.

Arrangements are made so that the President and other senior officers of government, as required, can be instantly reached by the Director of Central Intelligence or by their own intelligence officers in any emergency. Experience over the years has proved that this system really works. There was not a single instance during my service as Director when I failed to reach the President in a matter of minutes with any item of intelligence I felt was of immediate importance.

The CIA has also set up a Board of National Estimates within the Agency, on which sits a group of experts in intelligence analysis, both civilian and military. The board prepares initial drafts of most estimates, which are then coordinated with USIB representatives. To deal with highly technical subjects, such as Soviet missiles, aircraft or nuclear programs, competent technical subcommittees of USIB have been established. And, in certain cases, experts outside of government may be consulted.

Obviously, the procedure of preparing and coordinating an initial draft of an estimate, presenting it to the USIB, formulating the latter's final report along with any dissenting opinions and submitting it to the policymakers is time-consuming. There are times when "crash" estimates are needed. One of these occasions was the Suez crisis of November, 1956. I had left Washington to go to my voting place in New York State when I received early on election eve a telephone message from General Charles P. Cabell, Deputy Director of the CIA. He read to me a Soviet note that had just come over the wires. Bulganin was threatening London and Paris with missile attacks unless the British and French forces withdrew from Egypt. I asked General Cabell to call a meeting of the intelligence community and immediately flew back to Washington. The USIB met throughout the night, and early on election morning I took to President Eisenhower our agreed estimate of Soviet intentions and probable courses of action in this crisis.

The contents of this and other estimates are generally kept secret. However, the fact that this mechanism exists and can operate quickly should be a matter of public knowledge. It is an important cog in our national security machinery.

When, on October 22, 1962, President Kennedy addressed the nation on the secret Soviet build-up of intermediate-range missiles in Cuba, the intelligence community had already been receiving reports from agents and refugees indicating mysterious construction of some sort of missile bases in Cuba. It was a well-known fact that for some time past, Castro—or the Soviets purporting to be acting for Castro— had been installing a whole series of bases for ground-to-air missiles. These, however, were of short range, and their major purpose apparently was to deal with possible intruding aircraft. Since the reports received came largely from persons who had little technical knowledge of missile development, they did not permit a firm conclusion to be drawn as to whether all the missiles on which they were reporting were of the short-range type or whether something more sinister was involved.

The evidence that had been accumulated was sufficient, however, to alert the intelligence community to the need for a more scientific and precise analysis. Reconnaissance flights were resumed, and the concrete evidence was obtained on which the President based his report to the nation and his quarantine action. This required, of course, not only the most careful intelligence analysis but prompt intelligence judgments. As the President stated, the air reconnaissance established beyond a doubt that more than antiaircraft installations were being constructed on Cuban soil. This was a case, incidentally, in which it was obviously necessary to give publicity to intelligence conclusions. Khrushchev's subsequent statements and actions testified to their accuracy.

Here was another case where a "crash" estimate was required. Most of the estimating can be done on a more ordered basis, although there is usually a sense of urgency in the whole field of intelligence.

But whether as estimate has had weeks of analytical work behind it or is produced "overnight," years of training in the whole tradecraft of intelligence analysis are part and parcel of the final product. For example, in the Cuban case, the estimate could only have been produced quickly because of devoted work over many years by the highest qualified technicians in photoanalysis. These men and women had reached such competence from the study of earlier photographs of missile sites that what would be entirely unintelligible or subject to likely misinterpretation in the hands of the novice produced clear and reliable intelligence for the experts.

There must be intelligence analysis on each and every country where our interests may be affected, as well as in specified fields of particular intelligence interest; for example, the Soviet achievements in the fields of nuclear physics, ballistics, aerodynamics and space; also in industry, agriculture, and transportation. Naturally, the political, economic and social situations of many countries may also be of significance. I recall that once I had to have quickly a massive amount of information about Greenland. Within a matter of minutes, there was laid before me a study of the geography, geology, climate, peoples and history of that little-visited area.

All this is by no means just a question of automation, of filing away old reports and pushing the right buttons and getting the answers. Automation is a help and speeds up the process. But as we move further into the age of scientific achievement, the complicated machines and scientific-detection devices require the greatest sophistication on the part of the operators and analysts. Without this, our scientifically produced information as well as that furnished by the tools of espionage would be of little use. For it is the patient analyst who arranges, ponders, tries out alternate hypotheses and draws conclusions. What he is bringing to the task is the substantive background, the imagination and originality of the sound and careful scholar.

The analyst has sometimes been described as the man who takes forty-nine documents and from them produces a fiftieth. He does not do this by combining all the others, condensing and summarizing them, but by comparing them for their similarities and contradictions and shaking them down until he has sorted out what is probably true and significant, what is probably true but insignificant, and what is doubtful. He is, in a sense, finding out from the mass of unanalyzed information at hand, what we really know with some surety and what its value is, and what we don't know. He must bring to this task an impartiality that cannot be influenced by the fact that on the one hand lives may have been risked to procure the information, or that, on the other hand, the "customers" in the intelligence community will be more satisfied to receive full answers to their questions than the available fragments that only answer part of their questions.

A single report, for example, on a technical installation somewhere behind the Iron Curtain may have been entitled by the intelligence officer responsible for the area, "Production of Fighter-Bombers at Plant X." At headquarters, however, comparing this report with others on the same subject from a variety of sources, the analysts may find that some reference to metallurgical problems in the construction of a new rocket is the one valuable item in the whole report and that the main body of it, consisting of statistics on aircraft productions, is inaccurate or perhaps out of date. The latter part will therefore be shelved and the minute item on the rocket may alone find its way into that "fiftieth" document where it will be clearly ticketed as "untested" or "of unknown reliability," and will remain so designated until further information from other sources confirms the truth of it or shows it to be in error or possibly the figment of some agent's imagination.

There are knowable things which happen to be unknown. Sometimes they are easy, sometimes very difficult, to find out about. But there also are matters you cannot surely find out about at all. In such cases, if the requirement for a reasoned guess is high enough, we enter another phase of intelligence work—that of estimating. You make estimates not only about the knowable things that are not obvious, you make estimates also about those things which are literally unknowable, as we shall see.

Here is an unsung and perhaps unspectacular part of intelligence work, but I have often seen spectacular results emerge from it when our intelligence analysts are called upon to produce the estimate that the policymaker requires.

Some estimates are requested by senior policy officers of government to guide them in dealing with problems before them or to get an idea of how others may react to a particular line of action we may be considering. Others are prepared on a regularly scheduled basis, as, for example, the periodic reports on Soviet military and technical preparations. Before some estimates are prepared, a hurry-up call is sent to those who collect the intelligence to try to fill certain gaps in the information required for a complete analysis of a problem. Such gaps might be in the military or economic information available, or in our knowledge of the intentions of a particular government at a particular time.

Finally, estimates are often prepared because some member of the intelligence community feels that a certain situation requires attention. The cloud in the sky may be no bigger than a man's hand, but it may portend the storm; and it is the duty of intelligence to sound an alarm before a situation reaches crisis proportions. While the charge is sometimes made that intelligence has failed to warn of some crises, the press and outsiders do not know the number of times that it has given the necessary warning because this, again, is one of the sides of intelligence that is not advertised.

One general range of subjects that receives constant attention and very frequent, regular estimates is the development of what we call military hardware, particularly by the Soviet Union. This means Soviet programs and progress in missiles, nuclear warheads, nuclear submarines, advanced type of aircraft and anything that might approach a breakthrough in any of the sectors of this field, as well as in the field of space. This is one of the most difficult tasks which faces the intelligence estimator.

Here one has to deal with Soviet capabilities to produce a given system, the role assigned to the system by the military and its true priority in the whole military field. It is always difficult to predict how much emphasis will be given to any particular system until the research and development stage has been completed, the tests of effectiveness have been carried out and the factories have been given the order to proceed with actual production. As long as a Soviet system is still in its early stages, our estimates will stress capabilities and probable intentions; as hard facts become available, it is possible to give an estimate of the actual programming of the system.

In 1954, for example, there was evidence that the Soviet Union was producing long-range intercontinental heavy bombers comparable to our B-52s. At first, every indication, including the 1955 fly-by I have described, pointed to the conclusion that the Russians were adopting this weapon as a major element of their offensive strength and planned to produce heavy bombers as fast as their economy and technology permitted. An estimate of the build-up of this bomber force over the next few years was called for by the Defense Department and supplied by the intelligence community. It was based on knowledge of the Soviet aircraft-manufacturing industry and the types of aircraft under construction, and included projections concerning the future rate of build-up on the basis of existing production rates and expected expansion of industrial capacity. There was hard evidence of Soviet capability to produce bombers at a certain rate if they so desired. At the time of the estimate, the available evidence indicated that they did so desire, and intended to translate this capability into an actual program. All this led to speculation in this country as to a "bomber gap."

Naturally, intelligence kept a close watch on events. Production did not rise as rapidly as had seemed likely; evidence accumulated that the performance of the heavy bomber was less than satisfactory. At some point, probably about 1957, the Soviet leaders apparently decided to limit heavy bomber production drastically. The bomber gap never materialized. This became quite understandable, as evidence of progress in the Russian intercontinental missile program was then appearing and beginning to cause concern. Thus, while previous estimates of *capability* in bomber production remained valid, policy changes in the Soviet Union necessitated a new estimate on our part as to future development of the heavy bomber.

Intentions can be modified or policies reversed, and intelligence estimates dealing with them can rarely by unqualified. Witness how, just recently, our own intentions concerning the Skybolt missile have changed and how this must affect the calculations of Soviet intelligence.

The Soviet missile program, like that of the heavy bomber, had various vicissitudes. The Soviets saw early, probably earlier than we did, the significance of the missile as the weapon of the future and the potential psychological impact of space achievements. They saw this even before it was clear that a nuclear warhead could be so reduced in weight and size as to be deliverable over great distances by the big boosters which they correctly judged to be within the range of possibility. Given their geographical situation—their strategic requirements differ from ours—they soon realized that even a short- or medium-range missile would have great value in their program to dominate Europe.

The origins of the program go back to the end of World War II, when the Soviet Union, having carefully followed the progress made by the Germans with their V-1 and V-2 missiles, made every effort to gather together as much of the German developmental hardware and as many German rocket experts as they could get their hands on while they were conquering Eastern Germany. The Soviets also hired a considerable number of German experts in addition to those they seized and forcibly deported.

It would be a mistake, however, to credit their missile proficiency today largely to the Germans. The Russians themselves have a long history in this field and developed high competence quickly. They never took the Germans fully into their confidence but pumped them dry of knowledge, kept them a few years at the drawing boards and away from the testing areas, and then sent most of them back home. While these people proved to be a useful source of intelligence to the West, they had never been brought into contact with the actual Soviet development and could tell little beyond what they had themselves contributed.

In the early postwar years there was a good deal of skepticism in the United States about the future of guided missiles. One of the skeptics was Dr. Vannevar Bush, the outstanding head of our wartime Office of Scientific Research and Development, which coordinated the work of some 30,000 scientists, engineers and technicians. As late as 1949 he raised serious questions whether the guided missile could be "made to hit anything at the end of its flight"; he also felt its cost would be "astronomical." He added that as a means of carrying high explosives, "it is a fantastic proposal." He felt that in view of the cost of atomic bombs, we would not "trust them to a highly complex and possibly erratic carrier of inherent low precision."[2]

While there were some eminent men of science who differed from this view, it nevertheless was widely held. In the postwar years, before we had developed the thermonuclear bomb and the small but relatively powerful nuclear weapons, we failed to give the attention to the guided missile which, in the light of hindsight, we should have given it.

Another reason for this failure, and here intelligence enters into it, was the fact that in the first decade after the end of the war, we had inadequate information with regard to the Soviet missile program.

Drawing boards are silent, and short-range missiles make little commotion. As the techniques of science were put to work and the U-2 photographs became available after 1956, "hard" intelligence began to flow into the hands of the impatient estimators. Their impatience was understandable, for great pressure had been put on them by those in the Department of Defense concerned with our own missile programs and missile defenses. Planning in such a field takes years, and the Defense

---

[2] *Modern Arms and Free Men* (New York: Simon & Schuster Inc., 1949).

Department felt that this was a case in which it was justified in asking the intelligence community to project several years in advance the probable attainments of the Soviet program.

As in the earlier case of Soviet bomber production, the intelligence community, I am safe in saying, would be quite content if it were not called upon for such crystal-ball gazing. But since military planning requires estimates of this nature, the planners say to the intelligence officers: "If you won't give us some estimate as to the future, we will have to prepare it ourselves. You intelligence officers should really be in a better position to make the predictions than we are." For the intelligence service to deny this would be tantamount to saying it was not up to its job.

Thus early figures of Soviet missile production had to be developed on the basis of estimated production and development capabilities over a period in the future. Once again it was necessary to determine how the Soviet Union would allocate its total military effort. How much of it would go into missiles? How much into developing the nuclear potential? How much into the heavy bomber, as well as the fighter planes and ground-to-air defense to meet hostile bombers? How much into submarines? And, in general, how much into elements of attack and how much into those of defense?

It was due to this measure of incertitude during the late 1950s that the national debate over the so-called missile gap developed. Then, based on certain proven capabilities of the Soviets and on our view of their intentions and overall strategy, estimates were made as to the number of missiles and nuclear warheads which could be available and on launchers several years in the future.

There is no doubt that tests of Soviet missiles in 1957 and afterward showed a high competence in the ICBM field. Soviet shots of seven to eight thousand miles into the far Pacific were well advertised, as, of course, was the orbiting of the first *Sputnik*. Their testing in the intermediate fields must also have been gratifying to them. But would they use their bulky and somewhat awkward "first generation" ICBM, effective though it was, as the missile to deploy, or would they wait for a second or third generation? Were they in such a hurry to capitalize on a moment of possible missile superiority that they would sacrifice this to a more orderly program? The answer, in retrospect, seems to be that they chose the orderly program. As soon as this evidence appeared, the ICBM estimates, as in the case of the bombers, were revised downward.

Today, after the Cuba incident of October 1962, when Khrushchev did install "offensive" missiles in Cuba, one may well ask whether their recent actions do not indicate that they are in more of a hurry with their missile program. They were willing to take great risks to get some IRBM and MRBM bases in Cuba to create the equivalent, as a threat to us, of a considerable additional number of ICBM bases in the heartland of Russia.

In any event, the intelligence collected on Soviet missiles has been excellent as to the nature and quality of the potential threat. Our intelligence was also both good and timely as to Soviet production of high-thrust engines and the work on *Sputnik*. And all this intelligence spurred us to press forward with our own missile and space programs.

There is an area of intelligence estimating involving military hardware that is confusing to the uninitiated. On innumerable occasions during my period of work with the CIA, I was asked how the United States stood as compared to the Soviet Union in various respects. Were our bombers better? Did we have more missiles? How did we stand in the race for nuclear weapons? Here I had to explain that, as intelligence officers, we were not experts on American military weapons development. The job of the intelligence officer is to appraise the military strength of other countries, not that of the United States.

It is important, however, for our own policymakers to have the answers to such questions about comparative strengths. To meet this need, procedures were set up during the Eisenhower administrations to form net estimative groups. Intelligence officers were always members of the groups; the other members included experts having full knowledge of United States programs in the particular family of weapons where comparisons were sought—missiles, bombers, nuclear bombs and the like. Then net estimates would be produced indicating the relative position of the two countries and, wherever possible, where we would stand in a few years given our own existing programs and our estimate of those of the Soviet. This proved to be a most useful exercise.

When one turns from the military to the political field, the problems for the estimators are often even more complex. Analysis of human behavior and anticipation of human reactions can never be assigned to a computer, and they baffle the most clever analyst.

More than a decade ago, in the autumn of 1950, this country had to face in North Korea the difficult decision of whether or not to push forward to the Yalu River and reunite Korea. If we did so, would the

Chinese Communists answer with a direct attack? Or would they stay quiescent—if, for example, Korean rather than U.S. and UN troops formed the bulk of the advance, or if we did not disturb the Chinese sources of electric power in North Korea?

At that time, we had good intelligence as to the location and strength of the Chinese Communist forces on the far side of the Yalu. We had to estimate the intentions of Moscow and Peking. We were not in on their secret councils and decisions. In such cases, it is arrogant, as well as dangerous, for the intelligence officer to venture a firm opinion in the absence of telltale information on the positioning and moving of troops, the bringing up of strategic supplies and the like. I can speak with detachment about the 1950 Yalu estimates, for they were made just before I joined the CIA. The conclusions of the estimators were that it was a toss-up, but they leaned to the side that under certain circumstances the Chinese probably would not intervene. In fact, we just did not know what the Chinese Communists would do, and we did not know how far the Soviet Union would press them or agree to support them if they moved.

One cannot assume that a Communist leader will act or react as we would or that he will always be right in *his* estimates of our reactions. In Cuba, in October of 1962, Khrushchev presumably "estimated" that he could sneak his missiles into the island, plant them and camouflage them, and then, at a time of his own choosing, face the United States with a *fait accompli* which we would accept rather than risk war. Certainly here he misestimated—just as some on our side had misestimated that Khrushchev would not attempt to place offensive weapons in Cuba, right under our nose.

The role of intelligence in the early phases of the Cuban crisis of October, 1962, was the subject of a public report by the Preparedness Subcommittee of the Armed Services Committee of the Senate, under the chairmanship of Senator John Stennis of Mississippi. The subcommittee's main conclusion reads as follows: "Faulty evaluation and the predisposition of the Intelligence Community to the philosophical conviction that it would be incompatible with Soviet policy to introduce strategic missiles into Cuba resulted in intelligence judgments and evaluations which later proved to be erroneous."

This criticism of intelligence was directed to the period in September and early October, prior to the obtaining of adequate photography. Then there had been certain intelligence estimates to the general effect that it was unlikely intermediate-range missiles, i.e., missiles which could reach far into the United States, would be introduced into Cuba by the Soviets. There were some people, however, notably Mr. McCone, the Director of Central Intelligence, who had expressed at the time serious premonitions, but the intelligence community generally felt that Khrushchev would not risk a course of action so directly threatening to the United States and one which subsequent activities showed he was prepared to abandon abruptly in the face of strong American reaction. Cuba is yet another instance to warn us that one must be prepared for the Soviets to do the unexpected, the unusual, the shocking, confident in his own ability to retreat, as well as to advance, when the opposition gets too hot and also confident that he can make these retreats without seriously affecting his own domestic position. With complete control of the media of communication within his own country, he can explain away a retreat in Cuba as just another example of the "peaceful" posture of the Soviet Union.

In the preparation of estimates with regard to Soviet policy, their actions and reactions, it is always well to have among the estimators one or two persons who are designated to play the roles of the devil's advocate, who can advance all the reasons why a Khrushchev could take an unusual, dramatic or, as viewed from our own vantage point, even an unwise and unremunerative course of action. Of course, one would reach rather ridiculous conclusions, and certainly wrong conclusions in most cases, if one always came up with an estimate that the abnormal is what the Soviet Union will *probably* do. It is well, however, that the policymakers should be reminded from time to time that such abnormalities in Soviet action are not to be excluded.

If some of our own estimators went wrong in the Cuban affair, Khrushchev and his advisers committed an even more serious misestimate in apparently concluding that he could get away with this crude maneuver without a stern American rejoinder. Intelligence officers have to face the fact that whenever a dramatic event occurs in the foreign relations field—an event for which the pubic may not have been prepared—one can usually count on the cry going up, "Intelligence has failed again." The charge may at times be correct. But there are also many occasions when an event has been foreseen and correctly estimated but intelligence has been unable to advertise its success.

This was true of the Suez invasion of 1956. Here intelligence was well alerted as to what Israel and then Britain and France were likely to do. The public received the impression, however, that there had been an intelligence failure; statements were issued by U.S. officials to the effect that the country had not been given advance warning of the action. Our officials, of course, intended to imply only that the British and French and Israelis had failed to tell us what they were doing. In fact, United States intelligence had kept the government informed but, as usual, did not advertise its achievement.

Sputnik is another example. Here, despite the general impression to the contrary, the intelligence community predicted with great accuracy Soviet progress in space technology and the approximate time when their satellite would be orbited.

On other occasions the press and the public have been mistaken about the actual role of intelligence in certain situations. Having reached their conclusions about what the intelligence estimate must have been in the light of the official action taken, they have proceeded to attack the intelligence services, even though, in fact, no such estimate had been made.

Take, for example, the Bay of Pigs episode in 1961. Much of the American press assumed at the time that this action was predicated on a mistaken intelligence estimate to the effect that a landing would touch off a widespread and successful popular revolt in Cuba. Those who had worked, as I had, with the anti-Hitler underground behind the Nazi lines in France and Italy and in Germany itself during World War II and those who watched the tragedy of the Hungarian patriots in 1956 would have realized that spontaneous revolutions by unarmed people in this modern age are ineffective and often disastrous. While I have not commented on any details of the 1961 Cuban operation and do not propose to do so here, I repeat now what I have said publicly before: I know of no estimate that a spontaneous uprising of the unarmed population of Cuba would be touched off by the landing.

Clearly, our intelligence estimates, particularly in dealing with the Communists, must take into account not only the natural and the usual, but also the unusual, the brutal, the unexpected. Actions and reactions can no longer be estimated on the basis of what we might have done if we had been in Khrushchev's shoes because, as we have seen at the United Nations, he took off his shoes. Often Soviet moves seem to be influenced by the theories of Ivan Petrovich Pavlov, the famous Russian physiologist who induced certain reflexes in animals and then, by abruptly changing the treatment, reduced the animals to a state of confusion. The Pavlovian touch can be seen in Khrushchev's abrupt changes in attitude and action. The scuttling of the Paris Summit Metting in

1960, when he had for years known about the U-2, the surprise resumption of nuclear testing just at the time the nonaligned nations were assembling in Belgrade in 1961, even the famous shoe-thumping episode, were staged so that their shock effect would help produce the results he desired. He probably hoped for the same shock effects from the missiles in Cuba. Estimates on how Communist leaders act in a given situation should take this characteristic into account.

The willingness of a country to accept unpopularity in defense of its vital interests can be an element of strength. Often, because of our desire to be "loved," this element has been lacking in American foreign policy, but that does not mean that we should emulate the "shock" techniques of a Khrushchev.

Of course, one rarely has knowledge of all the factors bearing on the decisions of others. No one can predict with assurance the workings of the minds of the leaders whose actions make history. As a matter of fact, if we were to set out to estimate what our policies would be in a few years hence, we would soon be lost in a forest of uncertainty. And yet our estimators are called upon to decide what others will do. Unfortunately, the intelligence process of making estimates will never become an exact science.

But at least progress has been made in assembling the elements of a given situation in an orderly manner so as to assist our planners and policymakers. It is possible, often, to indicate a range of probabilities or possibilities and to isolate those factors which would influence Kremlin or Peking decisions. In any event, we have come a long way since Pearl Harbor and the somewhat haphazard system of intelligence analysis which prevailed at that time.

12

The Man on the Job

THE AMERICAN INTELLIGENCE OFFICER

The establishment of a permanent intelligence organization in the United States in 1947 resulted in the creation, for us, of a brand-new profession—the career intelligence officer. The profession is small, to be sure, but it still is a fact that this country is now offering to carefully selected young men and women the opportunity to make a lifetime career of intelligence work.

Intelligence officers were trained by the thousands during World War II, most of them to return to their civilian occupations when the war was over. At present the Army, Navy and Air Force maintain peacetime intelligence units which include civilians. For the most part the military personnel assigned to these units are on rotation and for limited tours of service. Until recently, a long tour of duty in intelligence was viewed by the ambitious military officer as a "graveyard" assignment, but this is no longer the case. However, the members of the armed forces who spend long tours in intelligence work are the exceptions.

From the day of its founding, the CIA has operated on the assumption that the majority of its employees are interested in a career and need and deserve the same guarantees and benefits which they would receive if in the Foreign Service or in the military. In turn, the CIA expects most of its career employees to enter its service with the intention of durable association. No more than other large public or private institutions can it afford to invest its resources of time and money in the training and apprenticeship of persons who separate before they have begun to make a contribution to the work at hand. It can, in fact, afford this even less than most organizations for one very special reason peculiar to the intelligence world—the maintenance of its security. A sizable turnover of short-term employees is dangerous because it means that working methods, identities of key personnel and certain projects in progress will have been exposed in some measure to persons not yet sufficiently indoctrinated in the habits of security to judge when they are talking out of turn and when they are not.

The very nature of a professional intelligence organization requires, then, that it recruit its personnel for the long pull, that it carefully screen candidates for jobs in order to determine ahead of time whether they are the kind of people who will be competent, suitable and satisfied, and that once such people are within the fold their careers can be developed to the mutual advantage of the government and the officer.

How is recruitment carried out in an intelligence agency, in particular in our own? The nature of the work which the candidate may be best suited to carry out is the controlling factor.

Initially you can't invite the prospect inside the plant and take him on a tour to show him how varied and rewarding a career in intelligence may be. Neither can you give him an illustrated booklet telling him all about the agency. Actually, the CIA does circulate a booklet about itself to inquiring job candidates, but this booklet cannot give information which would comfort the enemy or convey much enlightenment to the candidate. The employer wants to know everything about the candidate before employing him, but at that stage he cannot tell much about his organization or the job that awaits the applicant if he is selected.

Obviously in such a situation it is up to the employer to judge not only whether the candidate is suitable but whether he will be happy once he learns more fully what he is to do. The candidate must take on good faith the employer's assurances. And the only way the intelligence organization can give such assurances is to search as deeply as possible into the life and mind of the prospect, for his own benefit as well as the benefit of the organization.

Security investigations are a purely negative part of this process. They are rigorous, as they must be, but ninety-nine out of a hundred young Americans could pass a security investigation without difficulty. It is not hard to understand why an intelligence organization in these times cannot employ persons with close relatives behind the Iron Curtain, or persons who were at one time associated with Communist or other anti-American movements, or who in the past have displayed weaknesses in personal behavior or moral judgment. Finding out these things about a man is, however, relatively easy compared to finding out whether he is the right man for the intelligence profession.

The difficulty here is that the jobs in intelligence are manifold and there is room for many kinds of talent. And within any category of jobs many different kinds of men and women may succeed in different ways. There is likewise no fixed profile of personal characteristics which can be used in the selection of personnel for intelligence. But there are certain prerequisites without which, in all probability, the candidate will neither succeed nor be happy in the long run.

When I recently addressed a class of junior trainees at CIA I tried to list what I thought were the qualities of a good intelligence officer. These were:

Be perceptive about people
Be able to work well with others under difficult conditions
Learn to discern between fact and fiction
Be able to distinguish between essentials and nonessentials
Possess inquisitiveness
Have a large amount of ingenuity
Pay appropriate attention to detail
Be able to express ideas clearly, briefly and, very important, interestingly
Learn when to keep your mouth shut

I would add to this list certain other qualifications, desirable in a good intelligence officer, which have less to do with working ability than with attitudes and motives.

A good intelligence officer must have an understanding of other points of view, other ways of thinking and behaving, even if they are quite foreign to his own. Rigidity and closed-mindedness are qualities that do not spell a good future in intelligence.

An intelligence officer must not be overambitious or anxious for personal reward in the form of fame or fortune. These he is not likely to get in intelligence work. But he must bring to the task that intangible which is one of the most necessary characteristics of an intelligence officer—motivation. What motivates a man to devote himself to the craft of intelligence?

One way to answer the question is to look at some of the people who make up the ranks of American intelligence today and see how they got there. Here is a man, now a senior supervisor in CIA, who fought in the European Theater in World War II, stayed on for the occupation of Germany, was in Berlin during the airlift of 1948 and was finally returned stateside and discharged. He discovered after three months in his old job that the once attractive occupation of making money no longer satisfied him in a world of continuing international conflict, of which he had some knowledge thanks to his wartime and postwar service. He wanted to be closer to some front where he could feel he was "engaged," where he was dealing with the things he felt counted most.

Another man, a younger one, graduated from college in the early 1950s. He majored in government and international affairs. His father hoped he would go into the family business but the son didn't want to settle down to this routine—not just then. He wasn't really sure what he wanted to do but what interested him from the small glimpse he had of it in his college studies, and what stirred him every time he read the headlines, were the commitments and problems of the United States abroad and the Soviet challenge to our way of life. He went to Washington to look for a job, worked for a while in a branch of the government that had little to do with foreign affairs, and then finally found in intelligence what he was looking for.

Still a third man, from a small town in the Midwest, without a college education, was drafted, assigned eventually to a signals unit overseas, became fascinated with the Far East, witnessed the Chinese Communist attack on Quemoy, was returned stateside and discharged. Thanks to the training the Army gave him, he could have gone into electronics, or perhaps opened a television repair shop. Instead, he turned up one day at CIA offering his services and was assigned to an important communications job overseas.

What all these men had in common was an awareness of the conflict that exists in the world today, a conviction that the United States is involved in this conflict, that the peace and well-being of the world are endangered, and that it is worth trying to do something about these things.

What moved them is something more complicated than pure patriotism and something deeper than a mere longing for excitement. There is in the intelligence officer, whether he operates at home or abroad, a certain "front-line" mentality, a "first-line-of-defense" mentality. His awareness is sharpened because in his daily work he is almost continually confronted with evidences of the enemy in action. If the sense of adventure plays some role here, as it surely does, it is adventure with a large measure of concern for the public safety.

With this motivation, an alert, inquisitive and patriotic individual with an adequate education can be molded into a good intelligence officer. It is this complex "motivational" aspect of a man for which the intelligence service must probe in the prospective employee. Education, talent and the highest security clearances will not make him an intelligence officer if he does not have this motivation.

The charge has been leveled against CIA that it recruits almost exclusively from the so-called Ivy League colleges in the East with an overtone that possibly we have too many "softies" and possibly too many "liberals" for the tough job the CIA has to do. It is quite true that we have a considerable number of graduates from Eastern colleges. It is also true that in numbers of degrees (many of the CIA personnel have more than one degree) Harvard, Yale, Columbia and Princeton lead the list, but they are closely followed by Chicago, Illinois, Michigan, University of California, Stanford and MIT. It is interesting, however, to note that taking the approximately one hundred senior officers of the CIA, statistics show that these officers have degrees from sixty-one different universities, representing all parts of the country. It is, in fact, a highly heterogeneous group of men, representative of the entire Untied States, with a certain number of the men having postgraduate degrees from foreign universities.

Everyone who applies in writing or in person to CIA can be certain that his application will receive serious consideration. If there is no suitable position for which he could qualify, he is told so, as soon as the papers he has submitted are studied. If he seems to have some qualifications which recommend him for an existing opening, he will be invited for an interview. If the interviewer is favorably impressed and feels that the candidate seriously wishes to seek long-term employment with CIA and is not just seeking the "thrills" of what he thinks "espionage" work might bring, the process of testing begins.

The Korean war period caused a rapid expansion in CIA personnel, but the growth in recent years has been at a relatively low pace. There is a constant and recurring need for specialists and technicians to fill specific jobs requiring highly developed skills. In addition to these is the pressing need to recruit and train a cadre of young professional intelligence officers who possess the potential for executive leadership and who will eventually assume the responsibilities of senior intelligence officers and leaders of CIA. This is called the Junior Officer Training Program and its members, of an average age of about twenty-six, are Junior Officer Trainees (JOTs). They go through a series of training courses, at first general in nature, followed by others of increasing concentration on intelligence operations which prepare the JOTs for a specific type of work. This is followed by a trial period of on-the-job application of their training which determines eventual suitability for assignment. While in the training status, the JOT is carefully supervised by a training officer who looks at him as an individual in a continuing effort to place him on the job for which he is best suited. This pragmatic approach has proved itself in actual practice.

To find men of talent and promise, CIA does not rely solely, or even principally, on persons who apply to it for jobs. It goes out and looks for them on the campuses of colleges and universities all over the country. CIA does not do its hiring through the ordinary Civil Service mechanisms which serve as a clearinghouse for many parts of the government. It does, however, give its employees the same insurance and retirement benefits as are received under the Civil Service system, and its pay scale and its method of accruing annual leave and sick leave are the same.

CIA has been developing a Career Service plan with the aim, among other things, of charting out ahead of time for a foreseeable period of years various positions and posts to which an employee is to be assigned. The plan is based, as feasible, on the employee's own stated preferences, which are matched against the likelihood of openings suitable to the employee and on the supervisor's judgment of the employee's capabilities. Ambitious young men and women may sometimes dream up career plans for themselves which are not entirely practicable or which stem from a somewhat inflated estimate of their own capabilities. Agency programming helps to air such ambitions well in advance and to provide the employee with a realistic assessment of his future. Chiefly, however, the idea is to avoid arbitrary or makeshift assignments and to try to give some sense and continuity to the series of jobs which a man or woman may fill over a period of years.

Women in CIA undergo much the same training as men and can qualify for the same jobs, except that overseas assignments for women are more limited. One reason for this is the ingrained prejudice in many countries of the world against women as "managers" of men—in their jobs, that is. An agent brought up in this tradition may not feel comfortable taking orders from a woman, and we cannot change his mind for him in this regard. In World War II, American women shared risks in intelligence missions with men. Some of them parachuted into France as members of American jump teams who were sent in to support the French underground. While there is little reason to assign them today to jobs which endanger life and limb, many of them have served as members of intelligence units in hostile or "hardship" areas where for periods of years they have worked alongside the men, completely isolated from the amenities of modern life as they knew them at home.

An intelligence service, whether it be CIA or any other, will usually be made up of three broad categories of personnel, designated in popular intelligence parlance as the "operators," the "analysts," and the "support people." The latter, not directly concerned with the management of intelligence operations or the analysis of information, are the men who maintain the communications or attend to the administrative tasks of finance, personnel, supply, transportation, etc. A large part of the security force comes under this heading also. Some of the work of this department is often not too different from what it might be in any complex modern bureaucracy except that it must be done under conditions of maximum secrecy and with a full understanding of the machinery of an intelligence organization. Special tribute is due these hard-working men and women who, while subject to the same restrictions and discipline as the others, necessarily miss out on some of the excitement and challenge the others experience. Of course, those in communications work often man dangerous and vital posts abroad and constitute the very lifeline of an intelligence system, since information is useless unless it reaches headquarters speedily and safely.

The operators and the analysts are, respectively, those who gather and those who process information. The analytical process within an intelligence organization, ranging from the initial sifting and evaluating of information received to the preparation of high-level studies, calls primarily for a well-trained mind free of prejudice and immune to snap judgment. A man who is more interested in intellectual pursuits than in people, in observation and thought than in action, will make a better "analyst" than an "operator." For this reason, it is no surprise that people from the academic professions fill many of the analytical jobs. The "operator," or, as he is frequently called, the "case officer," is the field man, the collector of secret intelligence from agents. It is he who locates, recruits and handles the primary sources of information. The operators are drawn from everywhere. There is really no norm and no pattern. The main thing is that they be lively, curious, tireless and endowed with a keen sense for people.

People who try for intelligence jobs usually have a considerable background, as a result of their chosen studies, in international affairs, history or languages; not because they planned an intelligence career, but for the same reasons which would probably lead them to an intelligence career. However, the so-called "tradecraft" of intelligence is unique to a degree that there are few colleges which provide studies which automatically place one man in a more advantageous position than another. The only influence previous studies or experience has on a man's career in intelligence is to direct him more toward the analytical or the collection side, as the case may be, or more toward one geographical area of the world than another, or, if he is a technical expert, into some specialized area of intelligence. However, while the analyst may devote himself to one such area or topic for years, the "operator" usually will not, because his abilities in the craft itself are more important than any specialized topical or area knowledge. He can expect to be moved around many times in the course of his career. He gets this knowledge of the craft from the training schools of the intelligence service, from working as a junior officer with his peers, and finally from assignments in which he is more or less on his own.

Training schools in intelligence draw on many methods used in other professions in order to give the future intelligence officer not only knowledge, but experience and confidence. Intelligence, unlike many other professions, is not a business in which a few major or even small mistakes in the actual practice of the craft can be chalked up with a smile and wisecracks, such as "Back to the old drawing board." It has this in common with the military profession. Intelligence schools will give many courses about areas and languages that are not too dissimilar from university courses except for the emphasis on those things of chief concern to the intelligence officer. It will also give courses on the substance of intelligence itself, how intelligence services work, how information is analyzed, how reports are written, etc. But the guts of such training is the practical business of field operations, and to teach this intelligence schools draw on the practice of law schools in using the case method, and of the military in creating simulated "live" situations in which the trainee is expected to behave exactly as he would if he were on his own in a foreign country.

In the "case" method, past operations of American intelligence and of the intelligence services of other countries are studied. In order to confront the student with the exact circumstances and chronology of such operations, he is given replicas of files containing all the messages, reports, instructions, traffic between headquarters and outposts, agent materials, results of investigations and of surveillances in chronological order, so that he can see the day-to-day progress and conduct of the case, see it unfold before him like the rather complicated plot of a very long novel. Having the advantage of hindsight, he can see where mistakes were made, what the choices were, what was foreseen and not foreseen. The law student studying the briefs of the lawyers, the presentations of counsel for the plaintiff and the defense before the court, statements of witnesses, etc., can see in retrospect where one lawyer failed to ask a witness a telling question, where a summation to a jury failed to emphasize the most convincing evidence. Similarly, the student of intelligence, through a study of real cases in all their detail, will gradually begin to notice how the intelligence officer in a certain instance may have neglected to ask his agent a question which, as it later turned out, might have pointed to the latter's duplicity, how he forgot to give him a danger signal to use in an emergency, how a too complicated system of communicating between agents fouled up an important channel of information because one man simply couldn't remember what he was supposed to do in a certain situation. This study of cases particularly brings to light the human failures that mark the history of intelligence and implants in the young officer an appreciation of the many unpredictable elements which will play a role in his work and which it is his business to prepare for and to expect in every job to which he will later be assigned.

He will study in minute detail most of the famous cases in the history of modern intelligence, some of which we have had reason to cite in earlier pages, with equal attention to the reasons for success and the reasons for failure. How did Redl, Sorge and other noted spies of the past get away with it for so long, and what brought about their downfall? How could the Soviets have compartmentalized the segments of the *Rote Kapelle* or of the Canadian network so that the capture or defection of one member would not have brought the whole structure tumbling down?

In this pursuit of specific methodology, he also acquires a comparative knowledge of the strengths and weaknesses of the techniques favored by different national intelligence services. He will begin to see certain consistent national characteristics and aims displayed in these methods in somewhat the same fashion as the student of foreign policy or of warfare sees them in a study of nations at peace and at war. In some measure he will therefore learn what to expect from some of his future opponents.

The "live" situations in the training school are intended to achieve somewhat the same end as combat training with live ammunition. Pioneer work along these lines was done during World War II in the Army schools which trained prisoner-of-war interrogators. The interrogator-trainee was put up against a man who was dressed like an enemy officer or soldier, acted like one who had just been captured and spoke perfect German or Japanese. The latter, who had to be a good actor and was carefully chosen for his job, did everything possible to trick or mislead the interrogator in any of the hundred ways which we had experienced in real interrogation situations in Europe and the Far East. He refused to talk or he deluged the interrogator with a flood of inconsequential or confusing information. He was sullen or insolent or cringing. He might even threaten the interrogator. After a few sessions of this sort, the interrogator was a little better prepared to take on a real-life POW or pseudo defector and was not likely to be surprised by one.

This is the method essentially in use in intelligence training today. The situations are, or course, more complicated than those which confront an interrogator. Also, the intelligence school goes one step further in creating situations which can best be compared to the training of a psychoanalyst, who must first himself undergo analytical treatment in order to qualify fully as a healer of the mentally ill. The "live" situations in which the intelligence trainee is placed are not only those which he may someday meet as an intelligence officer. He must also play the role of the "agent" in them, not because he is likely to be an agent himself, but solely in order that he may begin to understand what it feels like to be inside the agent's skin and to develop greater sympathy and understanding—empathy would be the right word—for the practical and emotional predicament of the people who are going to work for him and take orders from him and often risk their lives for him.

The practical difficulties which a career in intelligence imposes upon a man and his family stem partly from the conditions of secrecy under which all covert intelligence work must be done. Every employee signs an oath which binds him not to divulge anything he learns or does in the course of his employment to any unauthorized person, and this is binding even after he may have left government employment. What this means is that an employee cannot discuss the substance of his daily work with his wife or his friends. Few have resigned or complained because of this particular constraint. Although it may sound like an almost paralyzing stricture to people who are unused to it, it does not work the hardship that may seem to be inherent in it. It may even have some social advantages in the sense that it forces people to be a little inventive, to develop hobbies and avocations and to take an interest in other things. I recall one outstanding intelligence officer (other than Rex Stout's Nero Wolfe) who made a hobby of orchids, others who wrote novels and mystery stories, still others, who, in their leisure, turned to music or painting. Most wives, after the honeymoon is over, easily tire of hearing their husbands talk about the office and the intricacies of their business, of the legal or governmental world in which they work.

The makeup of the personnel of CIA is as representative of all classes and places in America as any other branch of the government or any large business organization, and more so than many. Some of its members never attended college or never finished. Many are first-generation Americans, who often bring with them knowledge of the more unusual languages, though this is by no means the only reason why they might be employed.

An intelligence service in a free society is not only an institution in a democracy in that it is the creation of the Congress and subordinate to the executive; it also mirrors in its membership the society which it serves and inculcates in its officers the principle that the necessary strictures of secrecy make it all the more important that at all times the conduct and efficiency of its employees as public servants must be exemplary.

If CIA recruitment fails to equip the Agency with the best minds to keep the country's intelligence ahead of all its adversaries, including the Soviet Union, we are not properly taking advantage of the unique opportunities this country affords. Congress has appropriated adequate funds and has given CIA a comprehensive charter. The executive under four Presidents since its creation in 1947 has given CIA strong support. We have the greatest pool of human resources available to any country in the world—our 185 million people, our citizenry, come of almost every race of people on this globe. Furthermore, a hard core of highly skilled professionals from World War II days, both from the ranks of the OSS and from military intelligence work, have remained on or reenlisted in the CIA and furnish this country with a nucleus of experts, schooled in the hard experiences of wartime intelligence operations of every kind.

In building an intelligence service, it is clear that one needs a variety of people: the wise and discriminating analyzer and collator of the raw intelligence collected from all the quarters of the globe; the technicians to help produce, marshal and monitor all the scientific tools of intelligence collection; the staff officers, case officers and liaison officers to direct into proper channels the overall search for intelligence. Each of these varied tasks requires high skills and careful training.

THE AGENT

The intelligence officer engaged in covert intelligence collection described above is a career staff member of the intelligence service, an American citizen, on duty in a particular place, at home or abroad, acting on the instructions of his headquarters. He is a manager, a handler, a recruiter, also an on-the-spot evaluator of the product of his operatives. The man whom he locates, hires, trains and directs to collect information and whose work he judges is the agent. The agent, who may be of any nationality, may produce the information himself or he may have access to contacts and sources "in place" who supply him with information. His relationship with the intelligence service generally lasts as long as both parties find it satisfactory and rewarding.

If the staff intelligence officer succeeds in locating someone who is attractive to the intelligence service because of his knowledge or access to information, he must first ascertain on what basis the potential agent might be willing to work with him, or by what means he could be induced to do the job. If the agent offers his services, the intelligence officer does not have this problem, but he must still ascertain what brought the agent to him in order to understand him and handle him properly; he might, after all, have been sent by the opposition as a penetration.

As motives, ideological and patriotic convictions stand at the top of the list. The ideological volunteer, if he is sincere, is a man whose loyalty you need rarely question, as you must always question the loyalties of people who work chiefly for money or out of a desire for adventure and intrigue.

Actually, ideology is not the most accurate word for what we are describing, but we use it for want of a better one. Few people go through the analytical process of proving to themselves abstractly that one system of government is better than another. Few work out an intellectual justification or rationalization for treason as did Klaus Fuchs, who claimed that he could take an oath of allegiance to the British Crown and still pass British secrets to the Soviet Union because "I used my Marxian philosophy to establish in my mind two separate compartments." It is more likely that views and judgments will be based on feelings and on quite practical considerations. Officials in Communist bureaucracies who are not utterly blind to the workings of the state that employs them cannot fail to see that cynicism and power-grabbing prevail in high places and that the people are daily being duped with Marxist slogans and distortions of the truth. Communism is a system which deals harshly with all but its fanatical adherents and those who have found a way to profit from it. Every Communist country is full of people who have suffered at the hands of the state or are close to someone who has. Many such people, with only a slight nudge, may be willing to engage in espionage against a regime which they do not respect, against which they have grievances or about which they are disillusioned.

The man engaged in espionage on behalf of his own country is committing a patriotic act. The man who gives away or sells his own country's secrets is committing treason. Today we frequently encounter quite another situation, in which it is usually unjust to speak of treason. The internal political conditions of the Communist nations, as was once the case in the Fascist nations, have caused thousands to flee their homelands, either to save their own lives or because of their vigorous disapproval of the government in power. If an escapee aids his hosts in the country of adoption against the country he has fled, he can hardly be said to be committing treason as that term is generally used.

The ideological agent today usually does not consider himself treasonable in the sense that he is betraying his countrymen. He is motivated primarily by a desire to see the downfall of a hated regime. Since the United States is not imperialistic and makes the distinction of opposing Communist regimes rather than peoples of those countries, there can be a basic agreement in the aims of the ideological agent and the intelligence services of free states.

The more idealistic agent of this type will not engage in espionage lightly. He may at the outset prefer to join some kind of underground movement, if there is one, or perhaps to engage in the political activities of exiles which aim directly at unseating the tyranny which dominates his country.

During World War II, one of my best agents in Germany, whose information was of the utmost importance to the Allied war effort, never stopped trying to persuade me that he ought to be allowed to take part in the then growing underground effort to get rid of the Nazis. Every time I saw him, I had to point out to him that by doing this he would attract attention to himself and would only jeopardize his security, and that his ability to continue to get us much-needed information, what he was doing, was more valuable. It was obvious that he felt frustrated, that he wanted to get into the fight. He had another point, which was that his position after the war was over would be much better if he had helped bring down the Nazis. Nobody would make a hero of him for having supplied intelligence to the Allies. Unfortunately, he was right in this. Another anti-Nazi agent who collaborated with me at that time was willing to give every kind of information except the kind that might directly lead to loss of lives of his countrymen in combat. These are distinctions made by people of conscience.

Every intelligence service also makes use of people who work chiefly for money, or out of a love for adventure or intrigue. Some people thrive on clandestinity or deception for its own sake, deriving a certain perverse satisfaction from being the unknown movers of events. Among Communist conspirators one frequently finds this trait. People who knew Whittaker Chambers claim that there was a definite streak of this kind in him. In the upside-down world of espionage, one also finds men driven by a desire for power, for self-importance, which they could not satisfy in normal employments. The agent is often in on big things. He can make himself interesting and important to governments and sometimes gains access to astonishingly high places.

There is a fine passage in a World War I spy story of Somerset Maugham's about why a certain man had taken to spying. Maugham says:

> He did not think [Caypor] had become a spy merely for the money; he was a man of modest tastes. . . . It might be that he was one of those men who prefer devious ways to straight for some intricate pleasure they get in fooling their fellows . . . that he had turned spy . . . from a desire to score off the big-wigs who never even knew of his existence. It might be that it was vanity that had impelled him, a feeling that his talents had not received the recognition they merited, or just a puckish, impish desire to do mischief.[1]

[1] W. Somerset Maugham, *Ashenden; or, The British Agent* (Garden City, N.Y., Doubleday & Co., 1927).

What Maugham shows us here is, of course, a fact that every good writer and psychologist knows, and every good intelligence officer, too: motives are rarely pure and single, but most often mixed. The possibility of money and protection might often tip the scales for the person who is ideologically motivated but does not quite have the courage of his convictions. Some intelligence services feel it is important that even the ideological collaborator accept from time to time some money, or some kind of favor or gift, since this makes the agent somewhat beholden to the service; it seals the bond. Both Whittaker Chambers and Elizabeth Bentley told how the Soviets who were running the penetrations of the United States government during World War II went to great lengths to foist salaries or bonuses even on "dedicated" American Communists who were working for them. When the latter consistently fought the idea of accepting any sort of remuneration, the Soviets finally had their way by presenting them with expensive Christmas gifts, which couldn't be refused, such as oriental rugs—"a gift from the Soviet people in gratitude for their help," as Boris Bykov, a Soviet military attaché in Washington from 1936 to 1938, expressed it.[2]

Among the cases of people who will commit espionage for pay there are those who are in financial trouble—either debts they cannot meet or the misappropriation of government funds they have no way of replacing. Fearing discovery and unable to raise funds from any legitimate source, such a person may eventually turn to a foreign intelligence service with an offer of information, if it will pay him enough to rescue him. That crimes of "economic corruption" are frequent behind the Iron Curtain is evidenced by the particularly stringent measures taken by the men in the Kremlin to counter them, which I have already mentioned. A man who will try to extricate himself in this fashion from criminal prosecution contrives his own entrapment in espionage and is likely to serve the intelligence service well since he sees no other recourse. It can, after all, find ways to denounce him at any time to his own authorities.

Going down the scale of human behavior into the murky area of the psychoneuroses, one begins to encounter other kinds of motives, if they can be called that, which may induce a person to engage in espionage or to commit treason. Maladjusted persons nourishing grievances or seeking escape from their immediate environment may turn to

---

[2] Bykov made this remark to Whittaker Chambers, who quoted it in his book, *Witness* (New York: Random House, Inc., 1952).

treasonable acts if the means for doing so is in their hands, i.e., if they occupy positions which give them access to information of use to another power. Many minor cases of espionage especially turn out to involve persons who were not blackmailed or pressured, had no real ideological convictions, did not want money and were not adventurers in any normal sense of the word. They simply found some kind of twisted satisfaction in committing the act. Most of the cases, for example, of low-ranking members of American military establishments abroad who have crossed over to the enemy side fit into this category. To an unhappy misfit sitting in his barracks in West Germany, it may seem that the grass is greener in Communist East Germany or Czechoslovakia, which can be reached in a matter of hours. Frequently, such defectors will grab a handful of documents from military offices in which they are employed and will take them along in order to facilitate their welcome into a society they imagine will offer them a more satisfactory way of life than they previously managed to lead. A serious instance of such neurotically based treason was exposed by the flight behind the Iron Curtain in 1960 of two technicians from our highly sensitive National Security Agency, William H. Martin and Bernon F. Mitchell.

In the end, both the staff intelligence officer and the agent are needed to get the job done. Neither can manage without the other. Their relationship is unique in the professions. They are quite unlike buyer and seller, neither of whom need concern himself overmuch with the character and motives of the other, as long as business is done properly. Nor are they like employer and employee, although there may be occasional payment for goods and services.

Whatever his motives, and I have outlined some of them above, the agent initially must put himself in the hands of a stranger, the staff officer, about whom he knows very little, considering the delicacy of the work to be undertaken. He does know that the staff officer represents a foreign power in all its majesty and unreachability. The staff officer, on his side, must recognize that a large measure of his own authority derives from the fact that his country's flag is waving behind him. In addition, however, his own person must be such that it inspires confidence, trust and respect in the agent. After all, the agent's feelings about the capabilities of the intelligence system for which he may be risking his life will be based on the example he sees before him. A further complication is that the staff officer must manage to keep the agent's good-

will, so long as the agent does his job, whether he likes him personally
or not. Above all, he must fathom the agent's motives in order to protect himself and his government, to see that neither is exploited, swindled or harmed.

What counts in the end, to be sure, is the quality of the goods, i.e.,
intelligence, that the agent delivers. However, when human beings
rather than machines are involved in the collection process, the intangible and complex business of motives, loyalties and personalities plays
an enormous role in the success or failure of the whole enterprise.

13

Myths, Mishaps and Mischief-Makers

MYTHS

A number of major and minor myths have grown up during the last decade about CIA and the craft of intelligence itself as we practice it today. These myths are in part the creation of hostile propaganda of Communist origin; more often they are the product of imagination or guesswork, thriving on a lack of public enlightenment and on the suspicion any secret organization arouses. Sometimes these myths grow out of news stories purposely launched to "flush" out the facts. In such instances the bigger the exaggeration, the better the chance, so the writers think, of drawing a denial or correction or at least some answer other than "No comment," which for years has been, and I believe properly, the stock reply when the press calls on the CIA for information.

CIA MAKES POLICY

I have frequently been asked what "myth" about the CIA has been the most harmful. I have hesitated in answering, I admit, because there were several to choose from, but finally chose the accusation that CIA made foreign policy, often cut across the programs laid down by the President and the Secretary of State and interfered with what ambassadors and Foreign Service officers abroad were trying to do.

This charge is untrue but extremely hard to disprove without revealing classified information. It is all the harder to disprove because to some extent it is honestly believed, and at times has even been spread, by people in government who themselves were not "in the know."

The facts are that the CIA has never carried out any action of a political nature, given any support of any nature to any persons, potentates or movements, political or otherwise, without appropriate approval at a high political level in our government *outside the CIA*.

Here is an example of one of the recent myths of alleged political interference by CIA. The charge was spread abroad that the Agency secretly supported the OAS generals' plot against de Gaulle. This particular myth was a Communist plant, pure and simple. One of the first to launch it, on April 23, 1961, was a leftist Italian newspaper, *Il Paese (The Country)*, used from time to time as a trial balloon for Communist propaganda; then *Pravda* took it up and Tass sent it out to Europe and the Middle East, and the leftist press of Western Europe echoed it. Geneviève Tabouis, a well-known French writer who had a big following several decades ago, kept the propaganda mill going with three fantastic stories that gave Moscow new fuel. Meanwhile, highly reputable Western papers and columnists began repeating the rumors, and an aura of respectability was given to a story which was intended to discredit American policy in general and the CIA in particular.

In this, as well as in most such cases, there is absolutely no way to disprove such rumors. There is nothing to get your teeth into. It is only your word against the rumor market, and in this particular case high officials in the French Government did nothing to stop its spread.

A fresh and abounding group of myths about the CIA, each more fantastic than its predecessor, has been born out of the Bay of Pigs incident. A book published in May of 1964 contains a new crop of them.[1] The books is largely based on statements attributed to four brave and leading members of the Cuban brigade which went ashore at the Bay of Pigs. The responsibility for telling the story lies with Haynes Johnson, a Washington reporter. One particular bit of mythology about CIA in this book which particularly disturbed me relates to the myth I have been discussing—that CIA interferes with government policy.

[1] Haynes Johnson, *The Bay of Pigs* (New York: W. W. Norton & Co., Inc., 1964).

In describing the last days before the invasion force pushed off for Cuba, Johnson tells us about one of the American military trainers of the brigade coopted from the American military services—an officer known to the brigade members only as "Frank". I know Frank: he is an able officer, but here he was not involved in high policy matters. His job was to see to it that the brigade got good military training. As his knowledge of Spanish was vague and as the English of the brigade members with whom he was dealing was far from perfect, there was plenty of room for misunderstanding. From what Frank has recently said, I am prone to believe that this was all a misunderstanding which the Johnson book has built up into a grave incident seemingly only to discredit the CIA.

Here is the story according to the book. Shortly before the brigade left Nicaragua for Cuba, Frank called in two of the leaders of the brigade, Pepe and Oliva (they became two of the four co-sponsors of the book). Frank told them, so they are credited with saying, that "there were forces in the Administration trying to block the invasion and Frank might be ordered to stop it." If he receives such an order, he said he would secretly inform Pepe and Oliva. Pepe remembers Frank's next words this way.

> If this happens you come here and make some kind of show, as if
> you were putting us, the advisors, in prison, and you go ahead with
> the program as we have talked about it, and we will give you the
> whole plan, even if we are your prisoners.

This and certain related statements in the book have been widely blazoned in the American press as evidence that the CIA was preparing to thwart the orders of the President if he should have decided to call off the invasion.

This is totally false.

In the first place, Frank has denied the story.

In the second place, governing orders with respect to the brigade once it had left Puerto Cabezas would *not* have emanated from Nicaragua, Guatemala or from anyone in that area. They would have come from a command post located elsewhere which had direct contact with the brigade at sea and where the authority was *not* in Frank's hands.

Thirdly, at the time of Frank's alleged conversation with Pepe and Oliva, I know of no forces in the administration trying to block the action. True, no decision had been reached; the entire matter was before the President for decision.

Fourth, in addition to the control of the brigade exercised through the command post as I have mentioned, the brigade at all times after it set sail for Cuba and up to the time that it entered Cuban territorial waters could have been controlled by American naval forces.

Finally, shortly after this particular incident, the President of the United States on the eve of the landing gave the order to cancel the brigade's airstrike designed to immobilize Castro's aircraft, which might, and did, attack the incoming ships. The CIA, despite its deep apprehension of the effect of this order, responded immediately and loyally to the President's decision. The brigade's airstrike was canceled as it was on the point of taking off.

Here then is another myth which, if credited, could help to build up the utterly false theory that CIA stood ready to cross up high government policy.

Congressman Les Arends, who as ranking Republican member of the House Armed Services Subcommittee for CIA is well briefed on CIA doings, had this to say in a speech in the House of Representatives on March 26, 1964, regarding this myth of policy making.

> The statement has been made that CIA meddles in policy. This is an
> often heard allegation about the Agency, but the facts do not support
> it. CIA is an intelligence organization and takes its direction from the
> policymakers. The late President Kennedy commented on this in Oc-
> tober, 1963, when irresponsible sources were alleging that CIA was
> making policy in Vietnam.

Then he quoted what the President had said publicly in an answer to a question at a press conference as to whether CIA had meddled in our policy regarding Vietnam.

> I can find nothing, and I have looked through the record very
> carefully over the last nine months, and I could go back further, to
> indicate that the CIA has done anything but support policy. It does
> not create policy; it attempts to execute it in those areas where it
> has competence and responsibility. I can just assure you flatly that
> the CIA has not carried out independent activities but has operated
> under close control of the Director of Central Intelligence, operating

with the cooperation of the National Security Council and under
my instructions.

Another related myth is the charge that CIA always supports dictators.
This too has been subtly suggested in all manner of ways by Moscow
propaganda. Since CIA does *not* support Communists or fellow travelers, it must, in Moscow's view, support capitalistic warmongers, colo-
nialists, *et al.* There is nothing in between. Ergo it must be dictators
who are supported. And this myth has often been repeated in non-
Communist literature.

The President and the State Department set the lines of foreign
policies; they alone determine the course of conduct of all elements of
the government in all areas of foreign activity. Despite this fact of our
governmental life, the myth of mysterious and independent policies and
activities of the CIA persists, and, I fear, it is only as we get better
educated to the facts and less inclined to fall for divisive propaganda
that these myths will collapse of their own hollowness.

With the Soviets using their vast subversive machine to upset free institutions wherever they can, it is all very well to say that we should satisfy everybody's curiosity—including that of the Soviet—by
acknowledging each step taken in the effort to counter them, and tell
whom we are helping and why and where. But this is the best way to
lose the battle, and we should not be jockeyed or angered into answering these attacks, even if this means that troublesome myths persist.

THE SOVIET SUPER SPY

Nobody minds being portrayed as invincible. I imagine the Soviets
derive a good deal of satisfaction from the popular image of their intelligence officers and agents that exists in the minds of some Westerners.
The value of the image is that it tends to frighten the opponent.

If I seem to have lent any support to the myth of the Soviet super
spy in my earlier characterization of the Soviet intelligence officer, I
would like to remind the reader that I was then writing of his training,
his attitudes and his background rather than of his achievements. The
examples of Soviet failures are legion. Their great networks of the past,
often too large in size, eventually broke up or were exposed, both as a
result of the vigorous measures of Western counterintelligence and as a
result of their own internal weaknesses. Their best-trained officers make

technical slips, showing that they too are fallible. Often, in situations where there is no textbook answer, no time to get instructions from headquarters and when individual decision and initiative is required, the Soviet intelligence officer fails to meet the test.

Soviet training of both intelligence officers and agents tries to drill the wayward element out of intelligence work, but it cannot be done. Harry Houghton endangered his position by spending the extra money he earned from spying on real estate ventures. He wanted to amass a fortune. Vassall spent it on fancy clothes. Each lived beyond his regular income, and this was bound, sooner or later, to attract attention. Hayhanen, the associate of Colonel Abel, one of Moscow's best spies, was an alcoholic. He was bound eventually to break up, to talk—and he did. Stashinski, the murderer, on Soviet orders, of the two Ukrainian exile leaders fell in love with a German girl and came into conflict with his KGB bosses over this relationship. It was the main cause of his defection. The Soviets seem to have taken too little note of these weaknesses.

The Soviets cannot eliminate love and sex and greed from the scene. Since they use them as weapons to ensnare people, it is strange that they fail to recognize their power to disrupt carefully planned operations. A typical instance is described by Alexander Foote in telling of his Soviet military intelligence network during World War II.[2] Maria Schultz, a Soviet agent of long experience, was married to one Alfred Schultz, another old-line Soviet agent who was under arrest in China for espionage. In Switzerland Maria fell in love with a radio operator who had been assigned to work with her, divorced her husband at long range and married the operator. This bit of disloyalty dismayed her old servant, Lisa Brockel, so severely that out of chagrin the latter one day called up the British consulate in Lausanne and told the officer who answered the phone enough to endanger the whole Soviet network. Fortunately for the Soviets, her English was terrible, she was hysterical and the consulate thought she was just another crank.

Time and again the Soviets and satellites make serious psychological misjudgments in the people they solicit as agents. They underestimate the power of courage and honesty. Their cynical view of loyalties other than their own kind blinds them to the dominant motives of free people. A good illustration of this failing on their part was the case of the

---

[2] *Handbook for Spies* (New York: Doubleday & Co., Inc., 1949).

distinguished Rumanian businessman, V. C. Georgescu. In 1953, after his escape from Communist Rumania and after he had acquired American citizenship, he was approached by a Communist intelligence agent, acting under Soviet guidance, with a cruel attempt at blackmail. The agent, posing as a Secretary in the Rumanian legation, told Georgescu in so many words that if he would agree to perform certain intelligence tasks for Rumania, his two young sons, who were still being held in Rumania, would be released and returned to their parents. Otherwise he could never expect to see his sons again. Georgescu courageously refused any discussion of the subject. He threw the man out of his office and reported the full details to the United States authorities. The Communist diplomatic agent was expelled from the United States. The whole case received wide publicity so harmful to Rumania's relations with this country that the Rumanians finally sought to repair their damaged prestige by acceding to President Eisenhower's personal request for the release of the boys.

Soviet intelligence is often overconfident, overcomplicated and overestimated. The real danger lies not in the mythical capabilities of the Soviet spy, though some are highly competent, but in the magnitude of the Soviet intelligence effort, the money it spends, the number of people it employs, the lengths to which it is willing to go to achieve its ends and the losses it is willing and able to sustain.

WE AMERICANS ARE TOO NAÏVE AND TOO NEW AT THE JOB

Americans are usually proud, and rightly so, of the fact that the "conspiratorial" tendencies which seem to be natural and inbred in many other peoples tend to be missing from their characters and from the surroundings in which they live. The other side of the coin is that the American public, aware of this, frequently feels that both in our diplomacy and in our intelligence undertakings we are no match for the "wily foreigner." Foreigners likewise attribute to Americans a certain gullibility and naïveté. There are also other aspects of this same general notion. One is that the American official is a rather closed-minded do-gooder, a bit of a missionary, who butts into things he doesn't understand and insists on doing things his way. This is the "American" we see in Graham Greene's *The Quiet American*. *The Ugly American* gives us another angle of the same prejudice—lack of true understanding and appreciation of local conditions and of local peoples abroad. The number of best-sellers with this theme seems to show that it is a popular one and that we enjoy seeing our compatriots depicted as stupid people. It is little wonder then that such mischief-creating prejudices also find their way into the American and foreign criticisms of our operations abroad, including the intelligence service.

I would like to say first of all that I much prefer taking the raw material which we find in America—naïve, home-grown, even homespun—and training such a man to be a good intelligence officer, however long the process lasts, to seeking out people who are naturally devious, conspiratorial or wily, and trying to fit them into the intelligence system. The reader will have noted that when I described our norms for the potential intelligence officer in an earlier chapter, I did not include such traits among them. The recruiter does not look for slippery characters. He is much more likely to shun or reject them. The American intelligence officer is trained to work in intelligence as a profession, not as a way of life. The distinction is between his occupation and his private character.

Hand in hand with this preconception goes the attitude that American intelligence is young, hasn't had time to grow up, and can't possibly have produced a cadre of able officers in its brief existence who can match the work of older services, be they friendly or hostile ones. My answer to this is simple. We have seen nations such as Japan and Russia, who until the turn of this century were positively feudal, catch up with the technology of the twentieth century in one generation without going through the centuries-long evolution of Western societies. We have also seen that when a country has had its industry and technology devastated, as happened to Germany and to some extent France and Italy in World War II, it had a certain advantage when it began to reconstruct because it had lost the encumbrance of superannuated methods and equipment and there was no reason not to start with the latest and newest things.

American intelligence has been in precisely this position. During World War II it learned from the old-line services the results of centuries of experience. When the time came to found a permanent service here after the war, it was possible—indeed, imperative—to construct this organization along lines that would enable it to cope with contemporary problems and not with areas and conditions that had existed fifty years before. It is not important that American intelligence is young in years. What is important is that it is modern and not hidebound or tied to any outdated theories. I would point here above all to its ability to adapt the most modern instruments of technology to its purposes. In this it has been a daring pioneer.

SECRET INTELLIGENCE OPERATIONS ARE NOT IN THE AMERICAN TRADITION; IF ENGAGED IN, THEY SHOULD NEVER BE ACKNOWLEDGED

This is only in part a myth, and one that is on the wane. However, it is still true today that there are some Americans who are suspicious of any "secret" agency of government. Certainly that agency must assume the burden of proof that its claim to secrecy is reasonable and in the national interest.

Fortunately, there is a growing awareness of the dangers we face in the Cold War and that they cannot all be met by the usual tools of open diplomacy. And even those who regret the necessity for it are reconciling themselves to the fact that national security requires us to resort to secret intelligence operations. Interestingly enough, I have found little hesitation on the part of Congress to support and to finance our intelligence work with all its secrecy. In the very law setting up the CIA, Congress has enjoined the Agency to "protect intelligence sources and methods from unauthorized disclosure," but has provided none of the tools to accomplish this, outside of the CIA itself.

Naturally, when our intelligence operations go wrong and blow up in the press, there is bound to be criticism, and sometimes unjustified criticism. Intelligence operations are risky enterprises, and success can rarely be guaranteed. Since generally only the unsuccessful ones become advertised, the public gains the impression that the batting average of intelligence is much lower than is really the case.

The ability of the CIA to recruit year after year a select and very able group of our young college graduates shows that the hesitation of Americans about intelligence in general has not gone very deep in the younger generation. I have found that our young recruits have a growing appreciation of intelligence work as a career where they can make a real contribution to our national security. In my ten years with the Agency I recall only one case out of many hundreds where a man who had joined the Agency felt some scruples about the activities he was asked to carry on. In this case, he was given the option of either an honorable resignation or a transfer to some other branch of the work.

There was one sensational secret operation, now in the public domain, which did worry some people in this country as being "unlawful," namely the flights of the U-2 airplane. People know a good bit about espionage as it has been carried on from time immemorial. The illegal smuggling of agents with false papers, false identities and false pretenses across the frontiers of other countries is a tactic which the Soviets have employed against us so often that we are used to it. But to send an agent over another country, out of sight and sound, more than ten miles above its soil, with a camera seemed to shock because it was so novel. Yet such are the vagaries of international law that we can do nothing when Soviet ships approach within three miles of our shores and take all the pictures they like, and we could do the same to them if we liked.

If a spy intrudes on your territory, you catch him if you can and punish him according to your laws. That applies without regard to the means of conveyance he has taken to reach his destination—railroad, automobile, balloon or aircraft or, as my forebears used to say, by shanks' mare. Espionage is not tainted with any "legality." If the territory, territorial waters or air space of another country is violated, it is an illegal act. But it is, of course, a bit difficult for a country to deny any complicity when the mode of conveyance is an aircraft of new and highly sophisticated design and performance.

As I said at the outset, some of our fellow citizens don't want anything to do with espionage of any kind. Some prefer the old-fashioned kind, popularized in the spy thrillers. Some would concede that, if you are going to do it at all, it is best to use the system that will produce the best results and is most likely to secure the information we need.

The decision to proceed with the U-2 program was based on considerations deemed in 1955 to be vital to our national security. We required the information necessary to guide our various military programs and particularly our missile program. This we could not do if we had no knowledge of the Soviet missile program. Without a better basis than we then had for gauging the nature and extent of the threat to us from surprise nuclear missile attack, our very survival might be threatened. Self-preservation is an inherent right of sovereignty. Obviously, this is not a principle to be invoked frivolously.

In retrospect, I believe that most thoughtful Americans would have expected this country to act as it did in the situation we faced in the fifties, when the missile race was on in earnest and the U-2 flights were helping to keep us informed of Soviet progress.

And while I am discussing myths and misconceptions, I might tilt at another myth connected with the U-2, namely, that Khrushchev was shocked and surprised at it all. As a matter of fact, he had known for years about the flights, though his information in the early period was

not accurate in all respects. Diplomatic notes were exchanged and published well before May 1, 1960, the date of the U-2 failure, when
Khrushchev's tracking techniques had become more accurate. Still, since
he had been unable to do anything about the U-2, he did not wish to advertise the fact of his impotence to his own people, and he stopped sending protests.

His rage at the Paris Conference was feigned for a purpose. At the
time he saw no prospect of success at the conference on the subject of
Berlin. He was then in deep trouble with the Chinese Communists. Following his visit to President Eisenhower in the fall of 1959, he had been
unable to placate Mao during his stop at Peking en route back from the
United States. Furthermore, he was apprehensive that the Soviet people
would react too favorably to President Eisenhower's planned trip to the
U.S.S.R. in the summer of 1960. Influenced by all these considerations,
he decided to use the U-2 as a good excuse for torpedoing both the trip
and the conference.

There is evidence of long debate in the Praesidium during the first
two weeks of May, after the U-2 fell and before the date of the Paris
Conference. The question was, I believe, whether to push the U-2 issue
under the rug or use it to destroy the conference. There are also reports
that Khrushchev was asked why he had not mentioned the overflight
issue when he visited the President in 1959, more than six months before the U-2 came down. He is said to have remarked he didn't wish to
"disturb" the spirit of Camp David.

Finally, to conclude the U-2 discussion, I should deal with one other
myth, namely, that when Powers was downed on May 1, 1960, everybody should have kept their mouths shut and no admissions of any kind
should have been made, the theory being that you don't admit espionage.

It is quite true that there is an old tradition, and one which was excellent in its day and age, that you never talk about any espionage operations and that if a spy is caught, he is supposed to say nothing.

It does not always work out that way in the twentieth century.
The U-2 is a case in point. It is, of course, obvious that a large number of people had to know about the building of the plane, its real
purposes, its accomplishments over the five years of its useful life and
also the high authority under which the project had been initiated and
carried forward. In view of the unique nature of the project, its cost
and complexity, this proliferation of information was inevitable. It
could not be handled merely like the dispatch of a secret agent across

a frontier. Of course, all these people would have known that any denial by the executive was false. Sooner or later, certainly, this would have leaked out.

But even more serious than this is the question of the responsibility of government. For the executive to have taken the position that a subordinate had exercised authority on his own to mount and carry forward such an enterprise as the U-2 operation without higher sanction would have been tantamount to admission of irresponsibility in government and that the executive was not in control of actions by subordinates which could vitally affect our national policy. This would have been an intolerable position to take. Silence on the whole affair, which I do not believe could have been maintained, would have amounted to such an admission. The fact that both in the U-2 matter and in the Bay of Pigs affair the Chief Executive assumed responsibility for what was planned as a covert operation, but had been uncovered, was, I believe, both the right decision to take and the only decision that in the circumstances could have been justified. Of course, any subordinate of the executive, such as the Director of Central Intelligence, would have been ready to assume all or any responsibility in either of these affairs—even the responsibility of admitting irresponsibility if called upon to do so. In theory, this may have appealed to some. In actual practice, I believe it was quite unrealistic.

Today in the field of intelligence, many admissions are made, either tacitly or by deeds and actions, as well as in words. When the Soviet Union agreed to exchange Francis Powers for their spy, Colonel Rudolf Abel, they were admitting what he was and who he was, just as clearly as if they had published the facts in the newspaper.

Intelligence has come a long way since the good old days when everything could be shoved under the rug of silence.

CIA, THE BAD BOY OF GOVERNMENT

There are other kinds of myths, more of the spiteful or backbiting sort, that one sometimes hears in more restricted and "knowing" circles. I doubt if many readers outside Washington have ever even encountered them, and so I will deal with them only in passing. They have to do primarily with CIA's relations with other parts of our government, especially those with whom it works most closely. First of all, it is in the nature of people and institutions that any "upstart" is going to be somewhat frowned upon and its intrusions resented at first by the more well-established and traditional organizations. CIA had to prove itself and gain the respect of its elders by showing what it could do and by submitting its employees and its work to the test of time. It has, in my opinion, withstood this test and earned the respect of its fellows in government. It has, at the same time, not swallowed up the personnel, the property or the functions of any other agency, despite its reputed size and its reputed budget. The statement that there are American embassies where the CIA personnel outnumber the Foreign Service personnel is a rather typical troublemaking bit of malice, as is the one that the CIA personnel in embassies can do what they please. The Soviets, it is true, have many embassies where the intelligence personnel outnumber the diplomats, but we do not. The Soviet ambassador is himself sometimes an officer of the KGB. I have yet to hear of a case where the American ambassador was a CIA man.

The American ambassador is the commanding officer of all American officials in the country to which he is assigned, including any CIA personnel. This is subject, however, to the overriding authority of the President and the Secretary of State, who are responsible for the conduct of our foreign relations and decide how our policy should be carried out. It is they, of course, who instruct the ambassadors and determine the roles and mission of the various segments of our overseas missions, which often include AID, USIA, military, intelligence and other official personnel. There have been times when, under instructions of the State Department, the CIA has carried on certain operations which were not disclosed to the ambassador in the country in which the operations may have originated. This is the exception rather than the rule and generally happens only in a situation where an intelligence operation may be in part based in country A but more directly affects the situation in country B.

THE CIA AND THE FBI ARE AT LOGGERHEADS

This is one of the favorite myths. Nothing is more newsworthy than an internecine war between government agencies, and the press likes to tell us that these two organizations—the FBI working in the domestic field and the CIA in the foreign field—are literally knifing each other. As a matter of fact, one of the most satisfactory features of my work as Director of Central Intelligence was the close relationship established with

Mr. J. Edgar Hoover, particularly in the field of counterintelligence work. Each agency, of course, also furnished the other a mass of related positive intelligence material. Their respective areas and roles are clearly defined and conscientiously respected. The often-cited case of Col. Rudolph Abel is one where close cooperation between the two agencies paid off handsomely. This is only one instance of many where our information has been pooled and Soviet espionage operations have been checkmated, both at home and abroad.

CIA—THE INVISIBLE GOVERNMENT OF THE UNITED STATES

And now comes the latest and most horrendous myth of them all—that CIA and its cohorts in intelligence, particularly the military intelligence services, constitute the invisible government of these Untied States. Such is the thesis which two authors developed in 1964 for the edification of friend and foe alike, in some 350 pages of scuttlebutt.[3]

Mixing fact and fiction, accusing the intelligence services of spending some four billion dollars a year—a fantastic exaggeration—the authors pose as knight-errants of the press to kill once and for all the dragon of "secrecy" in government affairs. They purport to expose to the public and to the Kremlin and Mao the inner workings of intelligence, particularly in so-called "cold war" operations directed against Communism. In doing all this, they have also endeavored to surface to the world the names of intelligence and cold war operatives insofar as they have been able to uncover them.

But if one reads with care and perception what these authors have to say, you will see that they are trying to prove that the government of the United States itself has, from time to time, during the last four administrations, engaged, sometimes with success and sometimes without it, in certain operations, all approved at the highest level in government, to thwart the cold war tactics of Communism. In their "disclosures," they have offered to our antagonists the greatest propaganda bonanza since *Sputnik*. Fortunately, however, there are so many patent errors in what they say that neither Moscow nor Peking is likely to credit their story or believe that American correspondents could be so naïve as to publicize such secrets of government. Misunderstanding our system as

[3] David Wise and Thomas Ross, *The Invisible Government* (New York: Random House, Inc., 1964).

the Communists do and not appreciating the limitations on government to do anything about what is printed, they could not conceive that any government in its senses would allow monstrous violations of security to appear in public print unless this government had the sinister purpose of deceiving them.

The one thing these authors may well have demonstrated is this: under our system of government, there is precious little which can be kept secret and hence it is a myth that any "invisible government" exists.

LITERARY MYTHS—THE SPY IN FACT AND IN FICTION

The spy heroes of the novelists rarely exist in real life—either on our side of the Curtain or on the other. The staff intelligence officer, at least in time of peace, is hardly ever dispatched incognito or disguised into unfriendly territory on perilous or glamorous missions. Except for the Soviet illegal who is placed abroad for long periods of time, there is no reason for an intelligence service to risk the capture and interrogation of its own officers, thereby jeopardizing its agents and possibly exposing many of its operations.

There was little resemblance between the exploits of Ian Fleming's hero, the unique James Bond, in *On Her Majesty's Secret Service*, which I read with the greatest pleasure, and the retiring and cautious behavior of the Soviet spy in the United States, Colonel Rudolf Abel. The intelligence officer, as distinct from the agent, does not usually carry weapons, concealed cameras or coded messages sewed into the lining of his pants, or, for that matter, anything that would betray him if he should be waylaid. He cannot permit himself, as do the lucky heroes of spy novels, to become entangled with luscious females who approach him in bars or step out of closets, lightly clad, in hotel rooms. Such lures might have been sent by the opposition to compromise or trap him. Sex and hardheaded intelligence operations rarely mix well.

The Soviet "new look," which uses socialite spies, like Ivanov in London and Skrypov, mentioned in an earlier chapter, in Australia, represents an exception to this general rule. It may well be that the Soviets, having found pay dirt in the Profumo affair with its disruptive consequences, may see some advantages in using vice rings to aid blackmailing operations in later intelligence exploitation or merely to discredit persons in government positions in the Free World. This would fit in with general purposes of bringing such governments into disrepute with their own people. Certainly, from the intelligence angle, one would not expect to find items of intelligence passed via call girls to be of high reliability.

If there are dangers, tricks, plots, it is the agent who is personally involved in them, not the intelligence officer, whose duty it is to guide the agent safely. Even in the case of the agent and his own sources, the disciplines of intelligence today call for a talent for inconspicuousness that should rule out fancy living, affairs with questionable females and other such diversions. Alexander Foote, who worked for the Soviets in Switzerland, describes his first meeting during World War II with one of the most valuable agents of the Soviets. This was the man known by the code name Lucy, whose exploits I have already given.

> I arrived first and awaited with some curiosity the arrival of this
> agent who had his lines so deep into the innermost secrets of Hitler.
> A quiet, nondescript little man suddenly slipped into a chair at our
> table and sat down. It was "Lucy" himself. Anyone less like the spy
> of fiction it would be hard to imagine. Consequently he was exactly
> what was wanted for an agent in real life. Undistinguished looking,
> of medium height, aged about fifty, with his mild eyes blinking be-
> hind glasses, he looked exactly like almost anyone to be found in any
> suburban train anywhere in the world.[4]

Most spy romances and thrillers are written for audiences who wish to be entertained rather than educated in the business of intelligence. For the professional practitioner there is much that is exciting and engrossing in the techniques of espionage, but those untutored in the craft of intelligence would probably not find it so. And that part of actual espionage which is crucial—the successful recruitment of an important agent, the acquisition of critical information—for security reasons only finds its way into popular literature when it is seared with age.

A useful analogy is to the art of angling. In fact, I have found that good fishermen tend to make good intelligence officers. The fisherman's preparation for the catch, his consideration of the weather, the light, the currents, the depth of the water, the right bait or fly to use, the time of day to fish, the spot he chooses and the patience he shows are all a part of the art and essential to success. The moment the fish is hooked is the

---

[4] *Op. cit.*, p. 137.

moment of real excitement, which even the nonfisherman can appreciate. He would not be intrigued by all the preparations, although the fisherman is, because they are vital to his craft and without them the fish is not likely to be lured and landed.

I have always been intrigued by the fact that one of the greatest author-spies in history, Daniel Defoe, never wrote a word about espionage in his major novels. In the eyes of many, Defoe is accounted one of the professionals in the early history of British intelligence. He was not only a successful operative in his own right but later became the first chief of an organized British intelligence system, a fact which was not publicly known until many years after his death. His most famous literary works, of course, are *Robinson Crusoe, Moll Flanders* and *Journal of the Plague Year.* Try if you will to find even the slightest reference to spies or espionage in any of these books. No doubt Defoe carefully avoided writing about any actual espionage plots known to him because of political considerations and an ingrained sense of secrecy. But a man with his fertile mind could easily have invented what could have passed as a good spy story and projected it into another time and another setting. I cannot dispel the conviction altogether that he never did this because, having the inside view, he felt that for security reasons he could not give a true and full story of espionage as it was really practiced in his day, and as a novelist Defoe was above inventing something at variance with the craft.

An unusual writer on certain aspects of intelligence work is Joseph Conrad. I would venture to suggest that Conrad's Polish background is responsible for his native insights into the ways of conspiracy and the way of the spy. His own father was exiled and two of his uncles executed for their part in a plot against the Russians. The Poles have had long experience in conspiracy, as long as the Russians and, in great measure, thanks to Russian attempts to dominate them.

Being the kind of man he was, Conrad was not likely to tell a spy story for the sake of the adventure and the suspense. He was interested in the moral conflicts, in the baseness of men and their saving virtues. Conrad does not even exploit the inherent complexities of the spy stories he invents because it is not what primarily interests him.

The literature on intelligence which I find the most engrossing is of the Conrad type—stories that deal with the motivation of the spy, the informer, the traitor. Among these who have spied against their own country, there is the ideological spy, the conspiratorial spy, the venal spy and the entrapped spy. At different times in history one or the other of these motifs seems to dominate, and sometimes there is a combination of more than one motif. Klaus Fuchs was the typical ideological spy, Guy Burgess the conspiratorial type, the Swedish Colonel Stig Wennerstrom apparently was the venal spy, and William Vassall the typical case of entrapment—and finally there is the spy of fiction. And if at least we get pleasure in reading about him, let us keep him for such uses—even though he be a myth.

MISHAPS

In 1938, a Soviet intelligence officer working undercover in the United States sent a pair of pants to the cleaners. In one of the pockets, there was a batch of documents delivered by an agent employed in the Office of Naval Intelligence. It was not easy to press the pants with the documents in the pocket, so the pants presser removed them and in so doing brought to light one of the most flagrant cases of Soviet espionage in American experience up to that time. It was also one of the most flagrant instances of carelessness on the part of a trained intelligence officer on record. The officer, whose name was Gorin, was eventually returned to the Soviet Union, where he surely must have been shot for his sloppiness.

There have been some notorious cases of briefcases left behind in taxis or trains by people who should have known better. A sudden and inexplicable absent-mindedness can sometimes momentarily afflict a man who has been carefully trained in intelligence and security. But the gross mishap is usually not the fault of the intelligence officer. More often it results from the arbitrary or even the well-meaning behavior of outsiders who have no idea what the consequences of their acts may be, and from technical failures and from accidents.

The kind landlady of a rather busy roomer noticed that his spare pair of shoes was down at the heels. She took them to the cobbler's one day on her own. It was a favor. The cobbler removed the old heels and discovered that in each was a hollow compartment containing some strips of paper covered with writing. Of course he informed the police.

One of my most important German sources during my days in Switzerland in World War II almost had a serious mishap because his initials were in his hat. One evening he was dining alone with me in my house in Bern. My cook detected that we were speaking German. While we were enjoying her excellent food—she was a better cook than a spy—she slipped out of the kitchen, examined the source's hat and took down his initials. The next day, she reported to her Nazi contact the fact that a man, who from his speech was obviously German, had visited me and she gave his initials.

My source was the representative in Zurich of Admiral Canaris, head of German military intelligence. He frequently visited the German Legation in Bern. When he next called there, a couple of days after our dinner, two senior members of the legation, who had already seen the cook's report, took him aside and accused him of having contact with me. He was equal to the assault. Fixing the senior of them with his eyes, he sternly remarked that he had, in fact, been dining with me, that I was one of his chief sources of intelligence about Allied affairs and that if they ever mentioned this to anyone, he would see to it that they were immediately removed from the diplomatic service. He added that his contacts with me were known only to Admiral Canaris and at the highest levels in the German government. They humbly apologized to my friend and, as far as I know, they kept their mouths shut.

Everybody learned a lesson from this—I that my cook was a spy; my German contact that he should remove his initials from his hat; and all of us that attack is the best defense and that if agent A is working with agent B, one sometimes never knows until the day of judgment who, after all, is deceiving whom. It was, of course, a close shave, and only a courageous bluff saved the day. Fortunately, in this case my contact's bona fides was quickly established. The cook's activities eventually landed her in a Swiss jail.

The Sorge Communist network in Japan was broken in 1942 as the result of an action which was not intended to accomplish this end at all. In fact, the person who caused the mishap knew nothing about Sorge or his ring.

Early in 1941, the Japanese began rounding up native Communists on suspicion of espionage. One of these, a certain Ito Ritsu, who had nothing to do with espionage, pretended to cooperate with the police while under interrogation by naming a number of people as suspects who were basically harmless. One of those he named was a Mrs. Kitabayashi, who had once been Communist but had forsaken Communism while living in the United States and had become a Seventh-Day Adventist. In 1936, she had returned to Japan and sometime later had been approached by another Japanese Communist she had known in the United States, an artist by the name of Miyagi, who was a member of the Sorge ring. Miyagi had thus exposed himself to Mrs. Kitabayashi needlessly, it seems, since she, as a teacher of sewing, could not have had access to any information of interest to Sorge. Ritsu knew nothing of all this. He apparently denounced Mrs. Kitabayashi out of malice, to get her into trouble, because she had ceased being a Communist. When the police arrested Mrs. Kitabayashi, however, she gave away Miyagi. Miyagi in turn led to one of the highly placed sources of Sorge, Ozaki, and so it went until the entire ring was rounded up.

It is, of course, true that the larger a network is, with its many links and the need for communication between its various members, the greater are its chances of being discovered. Nevertheless, nothing that any of Sorge's very numerous and very active agents ever did aroused the attention of the police at any time. The officers who talked to Mrs. Kitabayashi couldn't have been more surprised when they were led, link after link, into one of the most effective espionage webs that ever existed. The discovery was purely the result of a mishap and one that no amount of careful planning could have avoided, except for just one precaution which the Soviets often failed to take: don't use anyone in espionage who ever was known as a party member.

The little slips or oversights which can give away the whole show may sometimes be the fault of the intelligence service itself, not of the officer handling the agent, but of the technicians who produce for the agent the materials necessary to his mission—the false bottom of a suitcase that comes apart under the rough handling of a customs officer, a formula for secret writing that doesn't quite work. Forged documents are perhaps the greatest pitfall. Every intelligence service collects and studies new documents from all over the world and the modifications in old ones in order to provide agents with documents that are "authentic" in every detail and up-to-date. But occasionally there is a slip that couldn't be helped and an observant border official, who sees hundreds of passports every day, may notice that the traveler's passport has a serial number that doesn't quite jibe with the date of issue, or a visa signed by a consul who just happened to drop dead two weeks before the date he was supposed to have signed it. Even the least imaginative border control officer knows that such discrepancies can point to only one thing. No one but the agent of an intelligence service would have the facilities working for him that are needed to produce such a document, which is artistically and technically perfect except in one unfortunate detail.

Then there is fate, the unexpected intervention of impersonal forces, accidents, natural calamities, man-made obstacles that weren't there the week before, or simply the perversity of inanimate things, the malfunctioning of machinery. An agent on a mission can drop dead of a heart attack, be hit by a truck or take the plane that crashes. This may end the mission or it may do more. In March 1941, Captain Ludwig von der Osten, who had just arrived in New York to take over the direction of a network of Nazi spies in the United States, was hit by a taxi while crossing Broadway at Forty-fifth Street and fatally injured. Although a quick-thinking accomplice managed to grab his briefcase and get away, a notebook found on von der Osten's body and various papers in his hotel room pointed to the fact that he was a German masquerading as a Spaniard and undoubtedly involved in espionage. When, shortly after the accident, postal censorship at Bermuda discovered a reference to the accident in some highly suspicious correspondence that had regularly been going from the United States to Spain, the FBI was able to get on the trail of the Nazi spy ring von der Osten was to manage. In March of 1942, their work culminated in the trial and conviction of Kurt F. Ludwig and eight associates. It was Ludwig who had been with von der Osten when the taxi hit him and who had been maintaining the secret correspondence with Nazi intelligence via Spain.

One windy night during the war a parachutist was dropped into France who was supposed to make contact with the French underground. He should have landed in an open field outside the town but was blown off course and landed instead in the middle of the audience at an open-air movie. It happened to be a special showing for the SS troops stationed nearby.

The now famous Berlin tunnel which went from West to East Berlin in order to reach and tap the Soviet communications lines in East Germany was a clever and relatively comfortable affair which had its own heating system, since Berlin winters are cold. The first time it snowed, a routine inspection aboveground showed, to the inspector's immense dismay, that the snow just about the tunnel was melting because of the heat coming up from underneath. In no time at all a beautiful path was going to appear in the snow going from West to East Berlin which any watchful policeman couldn't help but notice. He quickly reported what he had seen. The heat was turned off and in short order refrigeration devices were installed in the tunnel. Fortunately, it continued to snow and the path was quickly covered over. In all the complex and detailed planning that had gone into the design of this tunnel, this was something no one had anticipated. It was a near mishap in one of the most valuable and daring projects ever undertaken. Most intelligence operations have a limited span of usefulness—a tunnel, a U-2 and the like. This is assumed when the project starts. The difficult decision is when to taper off and when to stop.

The Soviets eventually did discover the Berlin communications tunnel and subsequently turned the East Berlin end of it into a public exhibit as proof to the East Germans of the long-advertised Soviet contention that the Allies only wanted to hold West Berlin because it was a convenient springboard for spying on the East. The Soviets set up an open-air beer-and-sausage stand near the spot so that the German burghers with their families could make a Sunday afternoon outing of their visit to the tunnel. This backfired, however, since the reaction of the visitors and the public in general was quite different from what the Soviets expected and wanted. Instead of shaking their fists at the West, the Germans got a good laugh at the Soviets because somebody had finally put something over on them and they were silly enough to boast of it. The beer-and-sausage establishment was quickly dismantled.

There is no single field of intelligence work in which the accidental mishap is more frequent or more frustrating than in communications. One of the best illustrations of this kind of mishap can be found in a well-known literary work which couldn't have less to do with intelligence. The reader will probably recall the incident in Thomas Hardy's *Tess of the d'Urbervilles* when the important message Tess slips under Angel Clare's door slides beneath the carpet that reaches close to the sill and is never recovered by the intended recipient, with grievous consequences for all.

Messages for agents are often put into "drops" or "caches," as places of concealment are called. These may be anywhere above ground or below ground, in buildings or out of doors. The Bolsheviks, like Dr. Bancroft, Franklin's secretary, used to prefer the hollow of a tree. Today there are safer and more devious contrivances by which means papers can be protected against weather and soil for long periods of time. In one case, the material was actually buried in the ground at a spot near the side of a road that had been used before successfully and was generally unfrequented day and night. On the occasion in question, the site was clear when the message was put into the ground, but when the agent came some days later to retrieve it, he found a mountain of dirt on top of it. In that short space of time between the placement and the arrival of the agent, the highway authorities had decided to widen the road and had begun to do so.

For obvious reasons, intelligence operations will often make use of public toilets as a place to cache messages. In some countries, they are about the only places where anyone can be sure of being absolutely alone. Even in such a place, luck can run against you. In one instance, the cleaning staff decided to convert one of the booths into a makeshift closet for their brooms, mops and buckets and they put a lock on the door. This was naturally the booth in which the message was hidden, and the conversion took place in the time between the placing of the message and the arrival of the agent to retrieve it.

In operations making use of radio communications, there can be a failure of the equipment on either the sending or receiving end. Communications making use of the mails can easily fail for at least ten good and bad reasons.

Often trains are late and a courier doesn't arrive in time to make contact with an agent who has been told not to wait longer than a certain time. To avoid this sort of accidental interruption of communications, most good operations have alternate or emergency plans which go into effect when the primary system fails, but here we begin to run into the problem of overload and overcomplexity, which is another quite distinct cause of mishaps. A person under some stress can commit just so much complex planning to memory, and will usually not have the plan written down because this is too dangerous. Or if he does have it written down, his notes may be so cryptic that he cannot decipher them when he needs to, even though when he wrote them down his shorthand seemed to be a clever and unmistakable reminder.

One of the simplest and oldest of all dodges used by intelligence in making arrangements for meetings calls for adding or subtracting days and hours from the time stipulated in a phone conversation or other message, just in case the enemy intercepts such a message. The agent has been told, let us say, to add one day and subtract two hours. Tuesday at eleven really means Wednesday at nine. When the agent was first dispatched, he knew this as well as his own name. No need to write it down in any form. Three months later, however, when he gets his first message calling him to a meeting, panic suddenly seizes him. Was it plus one day and minus two hours or was it minus one day and plus two hours? Or was it perhaps plus two days and minus one hour? Or was it . . . and so on. This is, of course, a very simple instance and hardly an example of the complex arrangements often in force.

Misunderstandings or forgetting of complex arrangements can lead to a delightful comedy of errors, especially when each party to a meeting or other arrangement tries to outguess or "second-guess" the other. The agent misses the meeting because he mixed up his pluses and minuses. The other party to the meeting was at the spot at the right time. When the agent didn't turn up, the other party imagined that the agent had mixed up his pluses and minuses and so tries to guess just how he mixed them up. He picks one of the four alternative combinations and goes to the spot again at that time. But he guessed the wrong combination. The agent in the meantime has remembered what was correct but it is too late because the correct day and hour have since rolled by. The two men fail to meet.

Mishaps, whatever their cause and nature, can be divided into those which reveal or "blow" the existence of an undercover operation to the enemy or to local authorities (which are not always identical) and those which simply cause the operation to fail or malfunction internally, such as when communications do not reach the right people but still do not fall into unfriendly hands. In either case, a major mishap, as in most of the cases I have been citing, may close off the operation for good or stall it for a very long time until the damage can be repaired, the communications re-established, etc.

Minor mishaps in intelligence have a nastiness all their own. One can never be quite certain whether they were damaging or not, and whether the operation should be continued or called off. Most of them have to do with losses of "cover," with partial or temporary exposure, instances where the inconspicuousness or anonymity of the agent is not maintained and he is spotted, even if only momentarily, as a person engaged in some kind of suspicious business, very possibly espionage. I might add that it will not help the execution of his task if the impression is made rather that he is a crook, swindler or smuggler.

Anyone who has ever traveled under another name knows that the greatest fear is not that you will forget your new identity while signing your name in the hotel register. It is rather that after you have just signed the register, someone will walk into the lobby whom you haven't seen for twenty years, come up to you, slap you on the back and say: "Jimmy Jones, you old so-and-so, where have you been all these years?"

Any operation involving the use of a person traveling temporarily or permanently under another name always risks the one-out-of-a-thousand chance that an accidental encounter will occur with someone who knew the agent when he had another identity. Perhaps the agent can talk or joke his way out of it. The trouble is that in today's spy-conscious world the first thing most people would think of is that espionage is the real explanation. If a great deal of work has gone into building up the new identity of the agent, such an accidental encounter might just ruin everything. The Soviet illegal is usually assigned to countries where the risk of such accidental encounter is minimal if not entirely nonexistent. Yet the following instance shows how the possibility always exists and how the Soviets, as well as the rest of us, have no way really of eliminating these risks entirely.

In the Houghton-Lonsdale case, as I have already stated, the American pair called Kroger who had been operating the radio transmitter were identified after their arrest as long-term Soviet agents who had previously been active in the United States. The FBI accomplished this identification on the basis of fingerprints. Just as the identification was completed their New York office received a phone call from a gentleman who described himself as a retired football coach. The week before, *Life* Magazine had shown a series of photographs of all the persons apprehended in the Lonsdale case. Thirty-five years ago, this gentleman told the FBI, he had been coaching at a large public high school in the Bronx. At that time a scrawny little fellow had tried out for the team, and he had never forgotten him. He had just seen Kroger's picture in *Life* and Kroger was that scrawny little fellow. He was absolutely certain of it. But his name wasn't Kroger, it was so-and-so. And the coach was right.

The Krogers had not tried to change their physical appearance at all. Kroger ran an open business in London of the kind that could have brought to him a variety of persons of all nationalities interested in collecting rare books. What was the chance that someone else, not necessarily the coach, who remembered him from that large public high school in the Bronx thirty or so years before would walk into his office one day in quest of a book and recognize him? Slight, but not impossible. The Soviets took the risk.

Minor mishaps may expose any of a number of elements that point to espionage. They may in many cases simply show that something out of the ordinary is going on, and whether this is interpreted as espionage and is therefore damaging depends in great measure on the innocence or sophistication of the beholder, whether he is, let us say, a policeman or a landlord or just a passerby. Frequently, they occur as a result of the agent practicing some of the known dodges and subterfuges of the professional agent which are, however, observed.

Once, somewhat unwisely perhaps, three men were sent to see a certain important personage who was occupying a suite of rooms on one of the upper floors of a hotel in a large European city. Each of them was a specialist and was needed for the opening gambit in this operation. They were not residing in the hotel or even in the country in question and were entirely unknown there. Many months later, after it had been established by other means of contact that this gentleman was willing to work with us, we sent one of the three original officers to see him. After some debate, it was decided less risky to send our officer to the hotel than to try to have the personage go out and meet us somewhere in the city, where few secure facilities were available to us. The officer had after all only been in the hotel once before, many months ago, and no one had the slightest means of knowing his business. Our man gave the number of the desired floor to the elevator operator. He was the only passenger. He looked over the operator, an old man and nondescript, and was sure he had never seen him before. But he was anxious to remember his face for the future because he would purposely avoid this particular chap and his elevator on his next few visits. Shortly before the elevator reached its destination, the old man turned around and looked at our man. "Oh, how are you?" he said. "I see you didn't bring your other two friends along today." Harmless? Probably, but you can never tell. The main point is that the officer was not so inconspicuous as he had thought. Elevator operators, like waiters and hotel people generally, remember faces. In certain countries, employees of this sort, bartenders, doormen, are police informants. Had he also guessed whom our man might be visiting? Had he guessed the nationality of our man, who spoke the local language well, but not perfectly? From his clothes, his manners? It is the very inconclusiveness of these minor mishaps which distinguishes them. The efficient intelligence service will take no chances after even the most minor mishap but will change its arrangements for contact and communications. It will even change the personnel on the job if it is the latter who are attracting attention.

MISCHIEF-MAKERS

One of the greatest sources of mischief for Western intelligence and diplomacy are the Soviet forgeries which I have already mentioned. Next in line I would rank the scurrilous propaganda which the Soviets manufacture, pretending to expose the personnel and methods of our intelligence services. To the perceptive Westerner these are generally funny, but their outlandishness is not likely to be perceived by the audience for whom they are intended. In their attempts to discredit American intelligence, the Soviets have produced for consumption behind the Iron Curtain and in neutral areas no end of books, pamphlets, press articles and radio programs branding our intelligence service as vicious. reactionary and warmongering, and its officers, including its Director, as gangsters and war criminals.

Such material is usually on the level of the lowest kind of war propaganda and revels in trumped-up stories and doctored pictures of atrocities. They have claimed that we torture people and have shown pictures of the instruments we use. More of such material has appeared in East Germany than elsewhere because the territory of East Germany has been most vulnerable to Western intelligence, and the Soviets rightly fear it and are anxious to frighten the East Germans away from any entanglements with the nefarious West.

One such work, published (in German) in East Berlin in 1959, is called *Allen's Gangsters in Action*. On its purple and yellow cover, it shows a partially unclad damsel who is wired with microphones and tape recorders and a miniature transmitter and antenna, all of which one would not see if she were fully clothed. Its general accuracy is attested to by the fact that it gives the address of CIA as "24 E-Street, Washington/N.Y." As anyone could have found out by consulting the Washington phone book, the older number was 2430 E, and, as we all know, the State of New York has not yet gobbled up the city of Washington.

A favorite tactic of such books is to accuse us of "brainwashing." As we know, the Soviets and the Red Chinese engage extensively in the brainwashing of prisoners of war in order to use the luckless victim for propaganda purposes. However, in accusing us of brainwashing, the Soviets are trying to explain to their own citizens how it was possible for a former Soviet or satellite national to speak up in the West against the Soviet system. They cannot admit that he was disillusioned and that he is acting freely and without prompting. They must insist that he was captured, even kidnapped, perhaps, and brainwashed, and has become the tool of the "imperialists" against his own will.

At times, however, though rarely, there is a touch of humor in the Soviet propaganda blasts. Some years ago, in a year-end summary of events and personalities which appeared in *Izvestia*, the well-known Soviet writer Ilya Ehrenburg devoted a few terse lines to me. He said in effect that if that spy Allen Dulles should ever pass through the "Pearly Gates" into Heaven, he would be found mining the clouds, shooting the stars and slaughtering the angels. I have found this a very useful introduction for public addresses where I attempted to outline the duties of the Director of Central Intelligence. Today Ilya Ehrenburg's writing generally seems to be more appreciated in the West than in Moscow.

Quite another kind of mischief-makers are the intelligence fabricators and swindlers. Among these there is the agent whose real sources "dry up" and who is therefore threatened with being put out of business. He knows what kind of information the intelligence service wants and he has its confidence. If he has no other means of livelihood and is not basically honest, it is understandable that he might come upon the idea of keeping the sources "alive" and functioning after they are really "dead" by writing their reports himself and fabricating their contents. Sooner or later the intelligence service will catch on, probably on the basis of internal evidence—errors in fact, discrepancies, an obvious paucity of hard data, a certain amount of embroidery that wasn't there before, even errors in style. Or the hoax might be exposed quite another way. The agent has to see his sources from time to time. When he does, he not only delivers to the intelligence service the information he collects, but writes a report on his meeting with the source, describing the circumstances of the meeting, the general welfare and state of mind of the source and many other matters which an intelligence service keeps track of. "Look here," says the intelligence officer to the agent. "You say you saw X on the twenty-fifth. That's very interesting, because we happen to know that he was out of the country all that week." This is not a pleasant moment for the intelligence officer if he is talking to a man who once did good work for him.

The intelligence swindler, as distinct from the real agent who has gone wrong, is a man who specializes in this sort of thing without ever having been a good agent for anybody. Like any other kind of swindler, he latches onto the latest racket except that his forte is to prey entirely on intelligence services, and from long experience he knows how to find their offices and how to get in the door. Fabricators and swindlers have always existed in the intelligence world, but the recent growth and significance of technical and scientific discoveries, especially their military applications, has afforded new and tempting fields for the swindlers. The weakness they could exploit was the lack of detailed scientific knowledge on the part of the intelligence officer. Although every modern service will train and brief its field officers as thoroughly as possible in scientific matters of concern to it, it clearly cannot turn every intelligence officer into a full-fledged physicist or chemist. The result is that many a good field officer may go for a neat offer of information and continue working with an agent until the specialists at home have had time to analyze the data and unhappily inform him that he is in the toils of a swindler.

Immediately after World War II, the most popular swindle by all odds exploited the new and world-wide interest in atomic energy. We were swamped with what we began to call "uranium salesmen." In all the capitals of Europe, they turned up with "samples" of U-235 and U-238, in tin canisters or wrapped in cotton and stuffed into pill bottles. Sometimes they offered to sell us large quantities of the precious stuff. Sometimes they claimed their samples came from the newly opened uranium mines of Czechoslovakia, where they had excellent sources who could keep us supplied with the latest research behind the Iron Curtain. There were many variations on the theme of uranium.

The chief characteristic and the chief giveaway of the swindler, as in most swindles, is the demand for cash on the line. First comes the tempting offer accompanied by the sample, then the demand for a large sum, after which the delivery of the main goods is to follow. Since no intelligence service allows its field officers to disburse more than token sums until the headquarters has reviewed a project in all detail, it is very rare that an intelligence service actually loses any money to a swindler. All it loses is time, but this is also precious, sometimes more precious than money. If the offer has any glimmer of truth to it and is not immediately recognizable as a swindle, an intelligence officer, for reasons I have already set forth many times, will try to hold on for a while in order to ascertain what he has. This can turn into a wasteful game of wits between the clever swindler and the intelligence officer, the latter refusing to let go entirely, the former fighting for all he is worth to put himself across and to parry all questions that would show him in his true light.

After uranium, there was a vogue in infrared, then came bogus information on missiles, and no doubt at this moment the swindlers are regrouping and working up reports on the Red Chinese development of a death-ray through the use of lasers. The logic here is that the Red Chinese are behind in H-bomb research and rather than go to the expense of catching up will devote their energy to lasers.

A more laborious and less easily identifiable kind of fabrication is that produced by what we call "paper mills." They turn out reports by the yard and do not depend on hot items as the swindlers do. Often their information is plausible, well reasoned and beautifully organized. There is only one fault with it. It doesn't come from the horse's mouth as claimed.

In their heyday, the paper mills exploited the situation created by the existence of the Iron Curtain and thrived in the late forties and early fifties when most of the Western services had not yet satisfactorily solved the problem of piercing the Curtain. During this period, many of the intelligentsia of Eastern Europe who had fled their homelands and had little hope of earning a living as refugees discovered that the intelligence services of the West were anxious to talk to them about conditions in the areas they had recently left behind them. The less scrupulous among them easily hit upon the idea of keeping these services supplied with what they needed. For this, of course, it was important to have "sources" behind the Iron Curtain, trusted friends in important jobs who had stayed behind, also clandestine means of staying in contact with these friends—couriers, smuggled correspondence, radio networks, etc. What made it difficult to prove that the information delivered was spurious was the fact that the authors were often well versed in the structure and habits of the governments and military organizations of their homelands and could take material from newspapers published behind the Curtain and from radio broadcasts and embroider on the information or interpret it with a good deal of art. Frequently, one had quite worthwhile information. The only trouble was it cost more than it was worth and didn't derive from the sources it claimed to derive from.

Shortly after World War II, a group of former military men who had escaped from one of the Balkan countries to the West promised us the plans of the latest postwar defenses on the Dalmatian coast, complete with harbor fortifications, missile ramps and the like. For this they wanted a good many thousands of dollars in gold. They agreed to show us a few samples of the papers before we paid up. These were supposed to be photocopies of official military drawings with the accompanying descriptive documents. They had allegedly procured the material from a trusted colleague, an officer who had remained behind and was now employed in the war ministry of an Iron Curtain country. In addition, there was a courier who knew the mountain passes, a brave man who had just come out with the plans and quickly returned home. He couldn't stay out in the West because his absence would be noted at home, and this was dangerous. If we wished to buy into this proposition, the courier would make a trip every month and the colleague in the war ministry would supply us with what we wanted on order.

The plans were beautiful. So were the documents. There was only one little flaw we noticed at the very first reading. Midway through one of the documents there was a statement that the new fortifications were being built by "slave" labor. Only an anti-Communist would use that term. There is, after all, no admitted slavery under Communism. Our military friends in their fervor had given themselves away. It was obvious that they themselves had drawn up the beautiful plans and documents in somebody's cellar in Munich. There was no brave courier and no friend in the war ministry, as they later admitted.

These paper mill products were usually cleverly conceived, well constructed and nicely attuned to the desires of the prospective purchasers and therefore almost impossible to reject on first glance. There was almost always a trained draftsman in the crowd, and the paper mill rarely failed to come up with elaborate and many-colored charts and tables drawn on a large scale showing networks of sources, subsources, letter drops, courier lines, safe houses and all the accouterments of professional espionage. As the result of a common drive on the part of the United States and other intelligence services, these mills have now for the most part been eliminated.

Cranks and crackpots run a close second after the fabricators as mischief-makers and time-wasters for the intelligence service. The reader would be amazed to know how many psychopaths and people with grudges and pet foibles and phobias manage to make connections with intelligence services all over the world and to tie them in knots, if only for relatively short periods of time. Again the intelligence service is vulnerable because of its standing need for information and because of the unpredictability of the quarter from which it might come.

Paranoia is by far the biggest cause of trouble. Since espionage is now in the atmosphere, it is no wonder that people with paranoid tendencies who have been disappointed in love or in business or who just don't like their neighbors will denounce their friends and foes and competitors, or even the local garbage man, as Soviet spies. During World War I, many German governesses employed by families on Long Island were denounced at one time or another and mostly for the same reason. They were seen raising and lowering their window shades at night, secretly signaling to German submarines which had surfaced offshore. Just what kind of significant information they could pass on to a submarine by lowering their shades once or twice was usually unclear, but then it is typical of paranoid delusions that there is a "bad man" close by, although it is never quite certain what he wants. Trained intelligence officers can frequently spot the crank by just this trait. There is usually very little positive substance to the crank's claim. The waiter at the "Esplanade" is spying for an Iron Curtain country. He was seen surreptitiously making notes in a corner after he had just taken overly long to serve two people who are employed in a government office. (He was probably adding up their bill.) It may later turn out that he had once accidentally spilled soup on the source, who was convinced he had done it on purpose.

Cranks and crackpots sometimes manage to wander from one intelligence service to another, and they can cause serious trouble if they are not spotted early in the game because they may have learned enough from the one experience to bring some substance to the next. A young and rather attractive girl once turned up in Switzerland with a story of her adventures behind the Iron Curtain and in West Germany and of her work in intelligence for both the Russians and one of the Allied services. Her story was long and took months to unravel. It was clear that she had been where she said she had been because she could name and describe the places and people and knew the languages of all the places. Most damning was her claim that certain Allied intelligence officers, including some Americans stationed in Germany, were working for the Soviets.

Our investigations eventually revealed that the girl had turned up as a refugee in Germany with information about the Soviets and the Poles, who had apparently employed her at one time in a purely clerical capacity. While the process of interrogation and checking was going on, she had come into contact with numerous Allied intelligence officers and had gotten to know their names. She apparently hoped for employment, but was finally turned down, since it was clear that she was a little wrong in the head. She next wandered into Switzerland, where she came to our attention. Her story by then had expanded and now included the men she had met in Germany, not in their true roles, but as actors in a great tale of espionage and duplicity. When she got through with us and went on to the next country, it is quite likely that the story got even bigger and that we who had just spoken with her also figured now as agents of the Soviets or worse. One of our people had the theory that the Russians had sent her to the West because, without any training at all, she was a perfect sabotage weapon. She could be guaranteed to waste the time of every intelligence service in Europe and prevent them from getting on with their more serious tasks.

14

The Role of Intelligence in the Cold War

Shortly before the Bolshevik revolution of October-November, 1917, a nationwide election was held in Russia for delegates to a Constituent Assembly, which was to choose the leaders of a new Russia.

This was the last, possibly the only, free vote the people of Russia ever had. Even under the chaotic conditions which prevailed in the fall of 1917 in war-torn Russia, about thirty-six million votes were cast for 707 Assembly seats. In this vote, the Bolsheviks received only about a quarter of the total and 175 seats. Unable either to control or intimidate the Assembly, Lenin dissolved it by brute force and the use of goon squads.

Here is Lenin's gloating judgment:

> Everything has turned out for the best. The dissolution of the Con-
> stituent Assembly means the complete and open repudiation of the
> democratic idea in favor of the dictatorship concept.
>
> This will be a valuable lesson.

And so it proved to be. The pattern was set for the techniques used in the destruction of freedom in other countries. Lenin here showed that a minority backed by illegal force could trample on a majority which re-lied on democratic methods.

It was some thirty years later before Communism felt it was strong enough to try these tactics outside of the area Russia had controlled in 1914, but as the war ended in 1945, Communism was on the march again. By then the Communists were consolidating their frontiers on the Elbe River deep in Western Europe, and had their forces of occupation and their subversive apparatus at work installing Communist regimes in Poland, Hungary, Rumania and Bulgaria. Shortly thereafter, they took over Czechoslovakia and had also begun their advance to the China Sea in the Far East.

A major part of the strategy of the Communists in the Cold War today is the secret penetration of free states. The means they use, the target countries they select and the soft areas in these targets are concealed as long as possible. They exploit secret weaknesses and vulnerabilities of opportunity and, in particular, endeavor to penetrate the military and security forces of the country under clandestine attack.

I include this issue—the most serious one we as a nation and the Free World face today—in a book on intelligence because intelligence has an important role to play here. The subversion campaigns of Communism generally start out using secret techniques and a secret apparatus. It is against them that our intelligence assets must be marshaled in good time and used as I shall indicate. Among the tasks assigned to intelligence, this is one that ranks in importance alongside those I have described: collecting information, counterintelligence, coordinating intelligence and producing the national estimates.

Of course, the whole range of Communist tactics in the Cold War is broader than the type of covert action and political subversion such as we have seen in Czechoslovakia and Cuba. It also includes: limited wars and wars by proxy, as in Korea and North Vietnam; guerilla wars, as in South Vietnam; civil wars, as in China; the use and abuse of their zones of "temporary" military occupation, as in the East European satellites and North Korea.

The Communists have not always succeeded, and this is due in no small measure to the employment of intelligence assets, not only of our own but also those of our friends and allies, including those of friendly governments under Communist attack. Their stooges took over power in Iran in 1953 and in Guatemala in 1954, and they were driven out. They tried to disrupt the Philippines and Malaya by guerrilla tactics, and they were defeated. They lavished arms deliveries on Egypt, Syria, Iraq and Indonesia, hoping these states would join the

Communist bloc, and so far they have had only a very modest return on these particular investments.

On the whole, however, they can look with satisfaction on what they have accomplished by subversion in the two decades since the Allied victory over Hitler and the Japanese war lords was assured in 1944. For it is wise to remember that the Communist program was well under way by the time of our peace talks with them at Yalta and Potsdam. Then they were thinking not of peace but of how they could use the common victory, and their zones of military occupation, for further Communist conquest.

In the last fifteen years, their progress has been considerably slowed down but by no means stopped. Beginning in 1947, they ran into a series of roadblocks: the United States stood firm in Greece, at Berlin and in Korea, and later on a broad front that reached to the Chinese offshore islands and Vietnam; helped by the Marshall Plan and other aid, Europe and Japan staged spectacular economic recoveries; Khrushchev and Mao Tse-tung were more and more divided on the tactics to pursue, although they remained in agreement on the basic objective of burying the Free World.

The Soviet policy of covert aggression rather than "hot" nuclear war had undergone considerable rethinking in the Kremlin following Stalin's demise and the revolution in Hungary. This policy was vigorously restated by Khrushchev under the general heading of "wars of liberation," in his speech of January 6, 1961. Here is how he outlined Communist power and Soviet tactics.

> Our epoch is the epoch of the triumph of Marxism-Leninism.
>
> Today . . . socialism is working for history, for the basic content of the contemporary historical process constitutes the establishment and consolidation of socialism on an international scale.
>
> The time is not far away when Marxism-Leninism will possess the minds of the majority of the world's population. What has been going on in the world in the 43 years since the triumph of the October Revolution completely confirms the scientific accuracy and vitality of the Leninist theory of the world socialist revolution.
>
> The colonial system of imperialism verges on complete disintegration, and imperialism is in a state of decline and crisis.

Later on in his speech, Khrushchev cited Cuba as the typical example of an uprising against United States imperialism. He then added:

Can such wars flare up in the future? They can. Can there be such uprisings? There can. But these are wars which are national uprisings. In other words, can conditions be created where a people will lose their patience and rise in arms? They can. What is the attitude of the Marxists toward such uprisings? A most positive one. These uprisings must not be identified with wars among states, with local wars, since in these uprisings the people are fighting for implementation of their right for self-determination, for independent social and national development. These are uprisings against rotten reactionary regimes, against the colonizers. The Communists fully support such just wars and march in the front rank with the peoples waging liberation struggles.

Now Communist parties are functioning in nearly 50 countries of these continents [Asia, Africa and Latin America]. This had broadened the sphere of influence of the Communist movement, given it a truly world-wide character.

Khrushchev concluded:

Comrades, we live at a splendid time: Communism has become the invincible force of our century.

This then is the charter of the Communists for world domination by world-wide subversion.

This country has been slow to arouse itself to the dangers we face from these tactics of Communism, which Khrushchev so clearly describes. Since Lenin's day this had always been a part of the Communist program. With Khrushchev, it became its major weapon in the foreign field.

In 1947, President Truman had proclaimed the doctrine which bears his name and applied it particularly to the then present danger of subversion facing Greece and Turkey. The doctrine, in effect provided that where a government felt that its "free institutions and national integrity" were threatened by Communist subversion and desired American aid, it would be our policy to give it. A decade later, this policy was restated in more precise language with respect to the countries of the Middle East in what became known as the Eisenhower Doctrine.

But these doctrines contained the general proviso that action would be taken if our aid were *sought* by the threatened state. Such was the case in Greece in 1947 and in Lebanon ten years later. In both instances, our assistance was invited in by a friendly government. The

Truman and Eisenhower doctrines did not cover, and possibly no officially proclaimed policy could cover, all the intricacies of situations where a country faces imminent Communist take-over and yet sends out no cry for help.

There have been occasions, as in Czechoslovakia in 1948, when the blow was sudden. Then there was no time for the democratic Czechs to send us an engraved invitation to help them meet that blow. We knew that the danger was there, that well over one-third of the Czech Parliament and several members of the Cabinet had Communist leanings and that the regime was seriously infiltrated, but the free Prague government of the day was overconfident of its own ability to resist. Between daylight and dusk, the Communists took over without firing a shot.

In Iran, a Mossadegh, and in Guatemala, an Arbenz had come to power through the usual processes of government and not by any Communist coup as in Czechoslovakia. Neither man at the time disclosed the intention of creating a Communist state. When this purpose became clear, support from outside was given to loyal anti-Communist elements in the respective countries—in the one case, to the Shah's supporters; in the other, to a group of Guatemalan patriots. In each case, the danger was successfully met. There again no invitation was extended by the *government* in power for outside help.

During Castro's take-over of Cuba, we were not asked by him for help to keep the Communists out; he was the very man who was bringing them in. Such crises show the danger of a slow infiltration by Communists and fellow travelers into a government where the last thing the infiltrators wish is outside intervention to check Communism.

What are we to do about these secret, underground creeping techniques such as were used to take over Czechoslovakia in 1948 and Cuba in recent years under the cloak of a Castro? Because Castro in one of his rambling and incoherent speeches has boasted about early Marxist views, the hindsight specialists are now saying that this should have been recognized years ago and action taken. Exactly what action, they do not specify except for those who advocate open military intervention. But thousands of the ablest Cubans, including political leaders, businessmen and the military, who worked hard to put Castro in and were risking their lives and futures to do so, did not suspect that they were installing a Communist regime. Today most of them are in exile or in jail.

First, I propose to review the main assets which the Kremlin can marshal for the tasks of subversion.

To simplify a complicated subject, I shall address myself solely to the apparatus of the U.S.S.R. Communist China, it is true, has similar aggressive purposes, but in the decade since they consolidated their position on the mainland, they have had neither the time nor the resources to develop a technique of subversion which is today comparable to that of the Soviet Union. This is one of the reasons for the emphasis they place on direct military action, as they have shown in the cases of Korea, Taiwan, India and Tibet. It may also be one of the reasons for the policy rift between them and the Soviet Union. The Chinese Communists feel that in their own case they cannot now rely on the more subtle techniques activated by the Soviets and would like to induce the latter to support direct military action. So far this is a policy that the Kremlin finds too dangerous, although it is not averse to using "nuclear blackmail" as a threat to intimidate other countries. In this way, Soviet military power influences the psychology of the situation, particularly in trying to soften up countries within easy range of its missiles and air force.

The first element of the Kremlin's nonmilitary apparatus of subversion is the galaxy of world-wide Communist parties. Here is Khrushchev's boast made as late as April, 1963:

> The international Communist movement has become the most influential political force of our epoch. . . . Before World War II Communist parties existed in 43 countries and counted in their ranks a total of 4,200,000 members. Today, Communist parties number 90 and the total number of their members exceeds 42,000,000.[1]

Most of these ninety parties are outside the Communist bloc but respond to discipline from the parent party in Moscow; in a limited but growing number of cases they look to the Chinese Communist party in Peking. Khrushchev's total numbers include only those who are actually party members and not the large numbers who vote the Communist ticket—where voting is permitted.

The most powerful Communist parties, numerically, outside the bloc are the parties in France, Italy, India and Indonesia, but numerical strength is not always the real test. For the purpose of subversion, an effective hard core of dedicated, disciplined members may be a more

[1] *New York Times*, April 22, 1963.

important factor than actual party membership. Wherever there is an organized Communist party, and that means in about every important country of the world and in many of the less important, there is generally a nucleus of dedicated Communists which can become an effective spearhead for subversive action.

Unfortunately, also, the local Communist parties in many countries have been able to establish themselves as the major party of protest against the regime in power. Thus they draw to their ranks, not necessarily as party members but as fellow travelers, on such issues as nationalism, anticolonialism "reform," and "ban the bomb," a large number of supporters who are really not Communists at all or who know and care little about Marxism and all its theories. At election time, the Communist party apparatus rallies together all these people and many others who are merely seeking a change and naïvely believe that the Communist party represents their best or sometimes their only vehicle for effecting a change.

Representatives of the Communist parties in the Free World regularly attend the party congresses in Moscow, of which the twenty-second was held in 1961. Here they are received as honored guests of the Congress and often are given special briefings. At the Twenty-first Party Congress held in 1959, the Communist delegates from Latin-American countries were given special attention. They were gathered together as a group and given secret guidance as to their methods of operation. At this particular time, to mislead the rest of the world and particularly the United States, they were told to play down Marxism and Communism and their relations with Moscow and to build their ranks by appealing to nationalism and using anti-American slogans. All this was not lost on Castro.

The Kremlin has always been willing, within bounds, to permit local Communist parties to take positions which differ from the official Moscow line. Sometimes this has been done by prearrangement with Moscow. On the other hand, the Kremlin has always had to cope with tendencies toward autonomy in other Communist parties. In recent years, as the Sino-Soviet schism has broadened, it has been increasingly difficult for the Kremlin to control the positions of all the other parties that were once subservient to it.

The tasks assigned by Moscow to Communist parties in Free World countries, and to the other elements of the Communist apparatus, are tailored to the estimated capabilities of the particular parties of "fronts," to the "softness" of the countries where they operate and to the general program of the Kremlin, i.e., the order of precedence for eventual take-over set by Moscow. For example, in the case of the Communist party of the U.S.A., where they have little hope of converting the country to Communism in the foreseeable future, the objectives assigned to the Party are relatively modest. They are told to stress propaganda against armaments in general and nuclear tests in particular; against American policy in Latin America; against NATO and our other alliances and our overseas bases. In England, it is much the same; "ban the bomb" is a chosen rallying theme. Such pacifist appeals are used to disguise real Soviet intentions and to soften the defenses of the Western world. In the spring of 1963, the "ban the bomb" movement achieved a level of unusual insidiousness through the publicity it achieved when it gave away the location of certain classified government centers prepared for emergency use in case of nuclear attack.

In countries where Communism has better prospects and more power, the horizon of objectives is raised. In France and Italy, the Communist party and its allies poll a vote which generally represent between 10 and 30 percent of the voters and, to the dismay of many who mistakenly believed that economic recovery alone would eliminate or at least weaken Communism, the Communists gained over a million votes in the Italian general elections of 1963. Here and in Indonesia, Japan, and in several countries of this hemisphere, as well as in Asia, the Communist parties take more aggressive positions. So far, in Africa, both north and south of the Sahara, Moscow's activities, both direct and through the local Communist parties, have been misconceived and ill-concealed.

Communist front organizations supplement the work of the local parties and are used as tools for reaching specialized objectives. For example, the Communists, through the World Federation of Trade Unions and its multiple branches, control the strongest labor organizations in many countries of the world—France, Italy and Indonesia in particular—and are able to manipulate significantly the unions in Japan, in many countries of this hemisphere and in certain countries of Africa and Southeast Asia, where trade unions are in their infancy. In the area of labor relations, the party makes particular use of its ability to "hitchhike" on popular local issues and to exploit them. Sometimes even where they do not actually control a union, well-organized and activist Communist minorities in unions can provide vocal and riotous leadership for mass demonstrations, and force a hesitant majority to engage in strikes and walk-outs, which are not openly attributable to any Communist initiative. Such activity at crucial times may paralyze the economy of an entire country.

Other Communist front organizations include the World Peace Congress, various youth organizations, women's organizations and organizations of specific professions. These they try to surround with a degree of respectability and to lure into membership the unsuspecting and the gullible, particularly on their "peace" and "ban the bomb" issues.

At various intervals, the Soviets at great expense to themselves have held "Youth Congresses," to which the youth of the world have been invited, but only the Communist youth get their way paid. Initially these meetings were held in areas behind the Iron Curtain—Moscow, East Berlin and Prague—but after that the Soviet managers of these affairs became bolder, and the last two meetings were held outside the bloc, first in Vienna and then in Helsinki. However, they found the climate of opinion so unfavorable in these capitals that they are now reconsidering whether to repeat the experiment.

Moscow's directing hand can help to guide and manipulate all these diverse assets of the Communist "presence" in a particular country through the State Security Service (KGB) personnel located in Soviet embassies and trade missions. The KGB, in addition to its regular intelligence function, can direct the activities of the local *apparat* set up in country X to promote a subversive program; they can act as Moscow's paymaster for the operations of the local party and fronts and keep Moscow advised of progress.

Valerian Zorin, who later became Soviet Ambassador to the UN, masterminded the Communist coup in Czechoslovakia in 1948 from within the Soviet embassy in Prague. The Soviet embassy in Havana was apparently also the center from which the early phases of the Communist infiltration of the Castro movement were directed.

Wherever possible Soviet tacticians will maneuver Communists or their sympathizers into key government positions and attempt to penetrate the target country's military and security structure with the idea of eventually taking them over. In the Allied Control Commissions which were set up in most of the Eastern European countries at the end of World War II immediately after the Germans had withdrawn, the Soviet contingents consisted largely of intelligence personnel. While the British and American representatives, specialists in military government and civil affairs, were trying to create some semblance of order and liberty and to restore the public utilities and the economy in devastated countries like Rumania and Hungary, their Soviet "colleagues" on the control commissions were spending their time working with reliable native Communists. Thus the conspiracies were organized which were shortly to emerge as "united fronts" dominated by Communists and supported by an efficient political police under KGB tutelage.

The vigor with which such tactics may be applied will depend as a general rule upon the circumstances in the target country: the extent of local unrest and of the local hostility to the regime in power, the capacity of the Soviet Union or Communist China to exploit latent vulnerabilities and suborn local political leaders and, finally, upon the strength of the Communist apparatus in the country in question.

Operating in countries which have recently obtained their freedom from colonial status, the Communist movement endeavors to present itself as the protector of the liberated peoples against their former colonial overlords. In support of these activities, promising young men and women from the target areas are invited to Moscow for education and indoctrination in the expectation that they may become the future Communist leaders in their homelands. Also they bring to the bloc for training in intelligence and subversion individuals of a different type who on their return will help to direct the local Communist party apparatus.

As a part of the *apparat*, Moscow also vigorously uses all the instrumentalities of its propaganda machine. In one year, according to the Soviet Ministry of Culture's report, the Soviets published and circulated approximately thirty million copies of books in various foreign languages. This literature is widely and cheaply distributed through local bookstores, made available in reading rooms and in their information and so-called cultural centers. In many countries throughout the world, they control newspapers and have penetrated and subsidized a large number of press outlets of various kinds which do not present themselves openly as Communist.

With some of the most powerful transmitting stations in the world, they beam their messages to practically every major area of the world. They step up their propaganda to the particular target areas which they consider to be the most vulnerable, and adjust it as their policy dictates. An organization known as the All Union Society for Cultural Relations Abroad, which poses as an independent organization but is strictly controlled by the Communist party of the Soviet Union, endeavors to establish cultural ties with foreign countries, supply Soviet films and arrange programs to be given by Soviet artists.

Then the foreign news agency of the Soviet Union, well known as Tass, a state-controlled enterprise, has offices in more than thirty major cities of the Free World. It adjusts its "news" to meet Soviet objectives in the recipient country. All these instruments of propaganda are part and parcel of what is called the *agitprop*.

These organizations and assets teamed together are, in a sense, Moscow's orchestra of subversion. Many of these instruments, and in some cases all of them, can be and are used under Moscow's careful supervision to bring pressure on any country they are seeking to subvert, or as a background to prepare for future subversion. They keep the orchestra playing, even to those countries like the United States, where the burying process, even by their estimation, is far removed.

Such is the apparatus of subversion we face today in the cold war the Communists have forced upon us, and I have added a glance at the history of the immediate past in dealing with it. To meet this threat we will need to mobilize assets and apply them vigorously at the points of greatest danger and in time—before a take-over, that is before a new Communist regime becomes firmly installed. Experience so far has indicated that once the Communist security services and the other elements of the *apparat* get their grip on a country, there are no more free elections, no way out.

Our assets against this threat are first of all our declared foreign policy, for which the State Department under the President has the burden of responsibility. Second, by our defense posture we can convince the Free World that we and our Allies are both strong enough and ready enough to meet the Soviet military challenge, and that we can protect, and are willing to protect, the free countries of the world, by force if need be; and that meanwhile we will aid them to build up their security against subversion. If the free countries feel that we are militarily weak or unready to act, they are not likely to stand firm against Communism.

A third element our intelligence service must help to provide: (1) it must give our own government timely information as to the Communist targets, that is to say, the countries which the Communists have put high on their schedule for subversive attack; (2) it must penetrate the vital elements of their subversive apparatus as it begins to attack target countries and furnish our government an analysis of the techniques in use and information on the persons being subverted or infiltrated into local governments; (3) it must, wherever possible, help to build up the local defenses against penetration by keeping target countries aware of the nature and extent of their peril and by assisting their internal security service wherever this can best be done, or possibly only be done, on a covert basis.

Many of the countries most seriously threatened do not have internal police or security services adequate to the task of obtaining timely warning of the peril of Communist subversion or of preparing to thwart it. For this they often need help, and they can only get it from a country like the United States, which has the resources and techniques to aid them. Many regimes in the countries whose security is threatened welcome this help and over the years have profited greatly from it. On the other hand, in some cases, especially in South America, a dictator has later taken over an internal security service previously trained to combat Communism and has diverted it into a kind of Gestapo to hunt down his local political opponents. This happened in Cuba under Batista.

Too often a threatened country feels that it can go it alone and sometimes too late awakens to the danger or comes quickly under the effective control of those who are promoting a Communist take-over. In these situations, there is no easy answer if no resistance is made and no call for help is sent out before the Communist apparatus crushes freedom. Often the apparatus uses its access to democratic processes, the ballot box and a parliamentary system, to infiltrate with what are called "popular front" governments. Then the mask falls away, the non-Communist participants in the coalition are eliminated and a Communist dictatorship has hold of the land and the secret police take over. Then it is too late indeed for protective action. Czechoslovakia is an example of this pattern.

Wherever we can, we must help to shore up both the will to resist and confidence in the ability to resist. By now we have had a good many years of experience in combating Communism. We know its techniques, we know a good many of the actual "operators" who run these attempts at take-over. Whenever we are given the opportunity to help, we should assist in building up the ability of threatened countries and do it long before the Communist penetration drives a country to the point of no return.

Fortunately for the Free World, because of the nature of the subversive activities in which the disparate Communist parties are engaged and the large numbers of untrained personnel involved, it is difficult for them to maintain adequate security and secrecy. It is revealing no secret to state that a very large number of the Communist parties and front organizations throughout the world have been penetrated. Often their plans and the personnel can be known. Dramatic information has already been published in regard to the effective work of the FBI in its penetration and neutralization of the Communist party of the United States and its various appendages.

Obviously it is somewhat more difficult for us to ferret out Communist activities in other parts of the Free World. But often it has been possible to achieve solid results which have prevented the Communists from reaching their objectives. Many Communist plots to subvert friendly governments have been discovered and thwarted. Local publicity in the early stages of a planned *Putsch*, pinpointing the plotters, tying them in to Moscow or Peking, has proved effective. This has been particularly useful in dealing with the bogus "front," "youth" and "peace" organizations of the Communists and their highly advertised meetings and congresses. Here a free press is also a great asset.

Formidable as is the Communist subversive apparatus, it is vulnerable to exposure and to vigorous attack. Furthermore, the Communists are in no position to push their program of take-over simultaneously in all quarters of the globe. They must pick and choose the areas which hold out the greatest promise to them. Meanwhile, on our side, there is much to be done, and a good deal is being done to shore up weaker countries and to keep them beyond the reach of the Communist grip. Certainly we must not limit ourselves to maintaining a defensive position and solely to reacting to Communist aggression. There have been instances where we have taken the initiative, where we have turned back the Communists, and there should be more. Apart from their problems at home and among themselves, many of their well-laid schemes to penetrate free countries have failed. After many frustrations in Central Africa, the Soviets appear to be regrouping and rethinking their prospects. Also as I have mentioned, their large investment in the Middle East and North Africa has been a bitter disappointment. In some areas of the world, they have found that lack of experience and ineptitude on the part of their envoys and agents, their parties and front organizations, have led to disaster. Their ignominious flight from the Congo in the early 1960s is a chapter in their history to put beside their earlier retreat from Albania.

The indigenous Communist parties are often torn between local issues and the policies of Communism. It is hard for them to shift as fast as Moscow does. One day they must bow down to a Stalin; then Khrushchev tells them Stalin is a bloodstained tyrant who betrayed the "ideals" of the Communist Revolution, and then Khrushchev is in turn purged. The Soviets preach Moscow's peaceful intentions and then have to explain the brutal crushing of the Hungarian patriots, just as earlier, in 1939, their strong appeal as an anti-Nazi force was dissipated overnight by Moscow's alliance with Hitler to destroy Poland, which Molotov called the "ugly duckling" of the Versailles Treaty.

As long as Khrushchev or his successors use their subversive assets to promote "wars of liberation"—which means to them any overt or covert action calculated to bring down a non-Communist regime—the West should be prepared to meet the threat. Where the tactic takes the form of open, hot or guerrilla warfare—as in Korea, Vietnam or Malaya—the West, on its side, can provide assistance openly in one fashion or another. But Western intelligence must play its role early in the struggle while subversive action is still in the plotting and organizational stage. To act, one must have the intelligence about the plot and the plotters and have ready the technical means, overt and covert, to meet them.

Of course, all actions of this nature undertaken by intelligence in this country must be coordinated at the level of policymaking and any action by an intelligence service must be within the framework of our own national objectives.

This country and our allies have a choice. We can either organize to meet the Communist program of subversion and vigorously oppose it as it insinuates itself into the governments and free institutions of countries unable to meet the danger alone, or we can supinely stand aside and say this is the affair of each imperiled country to deal with itself. We cannot guarantee success in every case. In Cuba, in North Vietnam and elsewhere, there have been failures; in many cases, many more than is publicly realized, there have been successes, some of major significance. But it may be premature to advertise these cases or the resources used.

Where Communism has achieved control of the governmental apparatus of a country, as it had, for a time, in Iran and Guatemala and it still has in Cuba and in Czechoslovakia, in East Germany, Hungary, Poland and the other Eastern satellites and in North Vietnam and North Korea, should we as a country shy away from the responsibility of continuing efforts to right the situation and to restore freedom of choice to the people? Are we worried that the charge be made that we, too, like Khrushchev, had our own policy of "wars of liberation"?

In answer to these two questions, I would point out that this issue, one important for our survival, has been forced upon us by Soviet action. In applying the rule of force instead of law in international conduct, the Communists have left us little choice except to take counteraction of some nature to meet their aggressive moves, whenever our vital interests are involved. Merely to appeal to their better nature and to invoke the rules of international law is of little use. We cannot safely stand by and permit the Communists with their "salami" tactics, so well advertised by Rakosi in Hungary, to take over the Free World slice by slice. Furthermore, we cannot safely take the view that once the Communists have "liberated" in Soviet style a piece of territory, this is then forever beyond the reach of corrective action.

If the people of a particular country, of their own free will, by free popular vote or referendum, should adopt a Communist form of government, that might present a different situation. So far this just has never happened. Neither Russia itself nor mainland China adopted Communism in this way. Certainly Poland, Hungary, Cuba and the others did not do so.

In the conduct of foreign relations, it must, of course, be recognized there are limits to the power of any nation. A country's enlightened self-interest, with all the facts taken into consideration, must guide its actions rather than any abstract principles, sound as they may be. No country could undertake as a matter of national policy to guarantee freedom to all the peoples of the world now under the dictatorship of Communism or any kind of dictatorship. We cannot go galloping around like Sir Galahad on his white charger, ridding the world of all its ills.

On the other hand, we cannot safely limit our response to the Communist strategy of take-over solely to those cases where we are invited in by a government still in power, or even to instances where a threatened country has first exhausted its own, possibly meager, resources in the "good fight" against Communism.

We ourselves must determine when, where and how to act, hopefully with the support of other leading Free World countries who may be in a position to help, keeping in mind the requirements of our own national security.

And as we reach our decisions and chart our courses of action in meeting Communist secret aggression, the intelligence services with their special techniques have an important role to play, new to this generation, perhaps, but nonetheless highly important to the success of the enterprise.

15

Security in a Free Society

Free peoples everywhere abhor government secrecy. There is something sinister and dangerous, they feel, when governments "shroud" their activities. It may be an entering wedge for the establishment of an autocratic form of rule, a coverup for their mistakes.

Hence it is difficult to persuade free people that it may be in the national interest, at times, to keep certain matters confidential, that their freedoms may eventually be endangered by too much talk about national defense measures and delicate diplomatic negotiations. After all, what a government, or the press, tells the people it also automatically tells its foes, and any person who through malice or carelessness gives away a secret may be betraying it to the Soviets just as clearly as if he secretly handed it to them. What good does it do to spend millions to protect ourselves against espionage if our secrets just leak away? Basically, I feel that government is one of the worst offenders.

Our founding fathers put the guarantee of freedom of the press in our Bill or Rights, and it became the First Amendment to the Constitution: "Congress shall make no law . . . abridging the freedoms of speech or of the press." As a result of this and other constitutional safeguards, it has generally been judged that although we have several espionage laws, we could not enact federal legislation comparable to that in effect in another great democracy, Great Britain. The

British Official Secrets Act provides penalties for the unauthorized disclosure of certain specified and classified information, and British legal procedures permit prosecutions without publicly disclosing classified information.

Our own method of dealing with security violations can, I think, be improved, and I propose later on to make certain suggestions in this regard. Anyone working in our own intelligence organizations in this country comes to realize, however, that it is necessary to plan with care and skill if he is to succeed in keeping his activities secret, and under present law he cannot expect much help from the courts in deterring those who would expose his activities. In fact, in my own experience in planning intelligence operations, I always considered, first, how the operation could be kept secret from the opponent and, second, how it could be kept from the press. Often the priority is reversed. For the intelligence officer in a free society this is one of the facts of life.

The question is whether we can improve our security system, consistent with the maintenance of our free way of life and a free press, and whether, on balance, it is worthwhile to try at least to limit our security lapses and indiscretions. I am persuaded that it is.

There are these important areas to be considered: *first*, the "giveaway," what is published with official approval; *second*, the "contrived leak," what is secretly passed out to the press by disgruntled or dissatisfied government officials who dislike a particular policy and feel that they must defend the position of their "service" against the encroachment of a rival service or the exponents of a conflicting policy; *third*, the "careless leaks." As a people, we talk too much; we like to show that we are in the know. Finally, there is the burning issue of the trustworthiness of the personnel who have access to classified information and the security of sensitive installations.

The recent disclosures of Pawel Monat, a Polish intelligence officer trained by Communist experts to carry on espionage in the United States, dramatize our national weaknesses. Colonel Monat was a high official of the Polish intelligence service before he was assigned to Washington in 1955 as military attaché. About three years later, in the spring of 1958, Monat returned to Poland, and after a year of further intelligence work there, and reflection on what he had experienced in the U.S.A., he decided to abandon his work and Communism. In 1959, he sought asylum in the United States through our embassy in Vienna.

Here are some of the things he has to say about espionage in the United States in his book *Spy in the U.S.*:

> America is a delightful country in which to carry out espionage. As a country it is rather ingenuous about keeping its secrets. . . . One of the weakest links in the nation's security . . . is the yearning friendliness of her people. . . . they crave public recognition. . . .
>
> I was able to find one American after another who seemed impelled—after a drink or two—to tell me things he might never have told his own wife.[1]

But it was obviously in published form that Monat found his most precious sources. "Americans," he says, "are not only careless and loquacious in their speech, they also give away far more than is good for them in public print."

Then he goes on to outline what he was able to get from one issue of *Aviation Weekly*, the "24th Annual Inventory of Air Power," which ran to 372 pages. "It would," he says, "have taken us months of work and thousands of dollars to agents to ferret out the facts one by one. . . . The magazine handed it all to us on a silver platter."

He pays tribute also to the publication *Missiles and Rockets* and very particularly to what he referred to as "house organs" of the Army, Navy, Air Force and Marines, which fight "the battle of interservice rivalry" in print, and to the stream of manuals and reports published by each of the services. Finally, he emphasizes the value to the Communist intelligence effort of "Congressional hearings on the defense budget," which he lists as one of his best sources.

"It must be extremely difficult," Monat adds, "for the U.S. military to try to defend the nation and its freedoms when the very sinews of its defenses are being exposed, day by day, to anybody who can read."

Douglass Cater, an eminent author and reporter, has frequently written about this whole problem and has dealt with it exhaustively and fairly. Describing the frustrations of both the Truman and the Eisenhower administration, he writes: "President Truman once claimed that '95% of our secret information has been published by newspapers and slick magazines' and argued that newsmen should withhold some

---

[1] *Spy in the U.S.* (New York: Harper & Row, Publishers, Inc., 1961).

information even when it had been made available to them by authorized government sources."[2] This, I feel, is a good deal to ask of any
newspaperman, though I have known of cases where reporters or their
editors on their own initiative have suppressed stories which they
deemed harmful to national security, or have sought advice as to the
sensitiveness of particular items.

In a press conference held by President Eisenhower in 1955, Cater
quotes the President as saying: "For some two years and three months I
have been plagued by inexplicable undiscovered leaks in this Government." Cater also refers to a statement by Secretary of Defense Charles
E. Wilson in which Wilson estimated that this country was giving away
military secrets to the Soviets that would be worth hundreds of millions
of dollars if we could learn the same type from them.

The intelligence community has been well aware of this problem,
and when he was Director of CIA Bedell Smith was so disturbed by
the situation that he decided to make a test. In 1951 he enlisted the
services of a group of able and qualified academicians from one of our
large universities for some summer work. He furnished them publications, news articles, hearings of the Congress, government releases,
monographs, speeches, all available to anyone for the asking. He then
commissioned them to determine what kind of an estimate of U.S. military capabilities the Soviets could put together from these unclassified
sources. Their conclusions indicated that in a few weeks of work by a
task force on this open literature our opponents could acquire important insight into many sectors of our national defense. In fact, when
the findings of the university analysts were circulated to President Truman and to other policymakers at the highest level, they were deemed
to be so accurate that the extra copies were ordered destroyed and the
few copies that were retained were given a high classification.

Is there any way to stop the giveaway? One large and important sector of this problem is within the control of the government and the Congress, that is, what the executive and legislative branches of government
publish or allow to be published, including particularly the publication
of Congressional hearings and investigations.

In this field, there is certainly evidence of influential Congressional
sentiment in favor of a move to curtail indiscriminate hand-outs.
On March 7, 1963, Representative George Mahon, a highly respected

[2] *The Fourth Branch of Government* (Boston: Houghton Mifflin Co., 1959).

member of the Congress and Chairman of the House Defense Appropriations Subcommittee, in a House speech widely reported in the press demanded an end to what he called an "outrageous and intolerable" situation. He asked that:

> . . . the President, the Vice President, and the Speaker of the House . . . undertake to coordinate a course of action for the purpose of halting the rapid erosion of our national intelligence effort. . . . Officials in Moscow, Peking, and Havana must applaud our stupidity in announcing publicly facts which they would gladly spend huge sums of money endeavoring to obtain. Responsibility on our part is urgently required.[3]

I, of course, recognize that in connection with appropriations and other legislation, particularly our defense budget, committees of the Congress need to receive a substantial amount of classified information from the executive. Does it necessarily follow that this must be published in great detail? It is often the intimate and technical details that are the most valuable to the potential enemy and of little interest to the public. I question whether, with respect to these technical details, there is a public "need to know."

It is also often said that Congress can't keep a secret. Past history belies this. The Manhattan Project, through which the atomic bomb was developed and billions of public funds spent, was a well-kept secret in a vital area of our national defense.

The reader may object that secrets can be kept in time of "hot" war but not under mere Cold War conditions. From almost ten years of experience in dealing with the Congress, I have found in my contacts with the subcommittees for the CIA of the Armed Services Committees of the House and Senate, and the Appropriations Committees of the two houses, that secrets can be kept and the needs of our legislative bodies met. In fact, I do not know of a single case of indiscretion that has resulted from telling these committees the most intimate details of CIA activities, and that included the secret of the U-2 plane. It is true, of course, that it is more difficult to preserve secrecy on matters which have to go before the entire Congress and receive its vote of approval. But it is not necessary to include intimate details of the kind that may have to be

---

[3] *Congressional Record*, March 7, 1963, p. 3549.

disclosed to certain Congressional committees by the Department of Defense in connection with its exhaustive budget presentations.

I would conclude that if this whole subject were discussed frankly and fully between the executive departments and the Congress, a method could be found for preventing the flow to hostile quarters of a part of the information which the adversary now obtains. There would still be a substantial trickle, to be sure, but not the great flood of information which is now made available. Is this not worth exploring?

A more difficult area is that of the press, periodicals and particularly service and technical journals. I recall the days when the intelligence community was perfecting plans for various technical devices to monitor Soviet missile testing and space operations. The technical journals exerted themselves to give the American public, and hence the Soviet Union, the details of radar screens and the like, which for geographic reasons, to be effective, had to be placed on the territory of friendly countries close to the Soviet Union. These countries were quite willing to cooperate as long as secrecy could be preserved. This whole vital operation was threatened by public disclosure, largely through our own technical journals, to the great embarrassment of our friends who were cooperating and whose position vis-à-vis the Soviets was complicated by the publication of speculations and rumors. Except for a small number of technically minded people, such disclosures added little to the welfare or happiness or even to the knowledge of the American people. Certainly this type of information did not fall in the "need to know" category for the American public.

Undoubtedly it is of the greatest importance in this nuclear missile age to keep the American people informed about our general military position in the world in ample detail. Of course we should have an informed public opinion, backed up with hard facts, authoritatively presented. There has been at times too much talk about bomber and missile gaps and the like. Personally, I am convinced that at no time has our military position been inferior to that of the Soviets. It is well that our people should know that and the Soviet Government, too. But what we don't really require is detailed information as to where every hardened missile site is located, exactly how many bombers or fighters we will have or the details of their performance.

If the giveaway is generally a result of our practice of conducting government in the open, both contrived and careless leaks can be attributed to interests and acts of special groups or individuals within the government. The contrived leak is the name I give to the spilling of information without the authority to do so, and it has occurred most often in the Defense Department and at times in the State Department. There have been cases where subordinate officers felt that their particular service or the policy which it is promoting was being unfairly handled by the press or even by higher officials of government because "all" the facts were not available to the press and public. It is, in effect, an appeal by subordinates, over the heads of superiors, to public opinion. This occurred once in connection with the transfer of major responsibility in the whole field of strategic missiles from the Army to the Air Force. At other times, information regarding State Department policies has been leaked by subordinates who disapproved of what was going on or by other agencies, generally the military, where there have been differences from State Department policy.

Douglass Cater cited a particularly disturbing leak of a private memorandum written by Secretary of State Rusk to Secretary of Defense McNamara, in which Rusk allegedly proposed that even "massive Soviet attacks on Europe should be met with conventional weapons." The story, Cater reports, "had not been based on the memorandum directly, only on an 'interpretation' of it, supplied by someone in the Air Force who was obviously hostile to the Secretary of State's position." He adds that it took an estimated one thousand man-hours of investigation before the Air Force general suspected of leaking the Rusk memorandum story could be identified, after which he was "exiled" to Maxwell Field, Alabama.

The careless leak, one not due to malice or plan, may be the result of someone talking thoughtlessly out of turn, perhaps encouraged by an astute reporter. By questioning enough people, the latter is often able to put together the true story of highly classified developments or programs in the making. All this is hard to deal with because reporters, who are directly or indirectly the beneficiaries of such leaks, refuse to disclose the sources, and it becomes almost impossible to obtain conclusive evidence as to who the guilty party, or parties, may be.

Very recently I found among the papers of my uncle Robert Lansing a most interesting letter and memorandum which President Woodrow Wilson, some fifty years ago, addressed to Lansing's pred-

ecessor as Secretary of State, William Jennings Bryan.[4] This proposed
a "panacea" to prevent leaks of secret White House–State Department correspondence. Here we see Wilson, who coined the phrase
"open covenants openly arrived at," trying, in his day, to deal with
the protection of our high-level diplomatic correspondence. The
"misplaced" memorandum enclosed with the President's letter of
February 8, 1915, to Secretary Bryan was obviously typed by Wilson
himself and has somewhat illegible interlineations in his own hand-
writing. Undoubtedly Bryan passed this correspondence on to Lansing when, a few months later, Lansing took over the office of
Secretary of State.

Woodrow Wilson, like all his successors, found only frustration
in this field of protecting secrets. He lived to see, in 1919, at the
Paris Peace Conference, one of the biggest diplomatic leaks of the
century. Then the terms of peace handed to the Germans at Versailles were, despite security precautions, prematurely leaked to the
American press.

Here is his 1915 plan to keep secrets from disclosure.

THE WHITE HOUSE
WASHINGTON

February 8, 1915

My dear Mr. Secretary:

Here is the memorandum of which I spoke to you some time ago
and which at that time I had misplaced. I submit these suggestions for
safeguarding the more important diplomatic proceedings for your consideration.

Cordially and faithfully yours,
Woodrow Wilson

enc.

Hon. William Jennings Bryan,
Secretary of State.

---

[4] The originals of Wilson's letter and memorandum together with certain other
Wilson-Lansing papers of World War I days, which the author recently found,
have been given to the Library of Princeton University.

MEMORANDUM.

One person to draft all despatches which it is thought wise to keep safe from publication.

One (and the same) stenographer to transcribe *all* such despatches and their ciphered or deciphered versions.

One (and the same) official to do *all* the enciphering and deciphering of such despatches.

No flimsies of such despatches; only one or two copies; a copy of the most important despatches to be sent to the President, to be returned for file always.

In brief, a single, clearly defined inner circle to handle these matters always, without variation of method or personnel, with the most carefully guarded exclusiveness, so that it may always be possible to fix the responsibility for a leak definitely and at once.

The only person outside this circle allowed even to *handle* such despatches nominated to be the head of the Index Bureau.

The despatches sent to the President to be sent always in sealed envelopes *to the White House*, never to the Executive Offices, where it is impossible to prevent their passing through several hands.

W. W.

February 12, 1915.

My dear Mr. President:

I have your letter of February 8th, enclosing your memorandum of suggestions for safeguarding the more important diplomatic proceedings of the Department. I think it will be entirely feasible to confine the matters of which you speak within the circle of you and myself and Mr. Davis, the Chief Clerk of the Department. Mr. Davis has been looking after these matters for some time, is familiar with the various ciphers used by the Department, and can also attend to the necessary typewriting of the despatches. This will seem to keep these most important matters within a very circumscribed circle which will be most advisable.

I am, my dear Mr. President,

Very sincerely yours,
(Sg) *William Jennings Bryan*

From the earliest days, the effort has continued to protect secrets by keeping knowledge of them to the fewest possible persons—on the "need to know" policy. Excellent as is this principle, it is generally defeated by the complications of modern governmental procedures. There are just too many who "need to know" or, what is worse, think that they do.

During my eleven years of service with the Central Intelligence Agency, I have attended scores of meetings at the highest level of government where a scene like the following has been enacted. It has been quite the same whether the administration has been Republican or Democratic. A high official of government, often the very highest, would come into a meeting brandishing a newspaper article and saying something like this: "Who is the so-and-so who leaked this? It was only a couple of days ago, here around this table, that a dozen of us reached this secret decision, and here it is all out in the press for our enemy's edification. This time we must find out who is responsible and string him to the nearest lamppost. We can't run a government on this basis anymore. This thing must stop. Investigate and report and this time get us some results. I don't propose to tolerate this sort of thing in this administration any further."

And then the wheels start to move. A committee on security whips into action; the FBI may be called in if it is surmised that a violation of a Federal statute is involved. In due course, the investigation comes up with the following results.

It is found that the particular decision of government which leaked out was set down in a secret or top secret memorandum of which, initially, there were perhaps a dozen copies for distribution to the various departments, agencies and bureaus of government which might be involved, on a strict "need to know" basis. Several hundred people then had access to this memorandum, because it was reproduced in multiple copies by department heads for the information of their subordinates. Messages also might have been sent to officials in various parts of the world where action might be required. When such an investigation has been concluded, it is often established that anywhere from five hundred to a thousand people might have seen the document, or heard of its contents and have talked about it to X, Y and Z. No official will ever admit a violation of security was involved in this process, and no newspaperman or publicist will ever give away a source.

After the investigation is closed, the verdict is that the offense has been committed by a person or persons unknown and undetectable. Somewhere in the course of this proceeding, the Director of Central Intelligence is generally reminded that the law setting up the CIA provides that it shall be the duty of the Director of Central Intelligence to "protect intelligence sources and methods from unauthorized disclosure." He is then asked what is being done to carry out the mandate of the law.

His reply generally is that the law has given him no investigative authority outside of his own agency and, in fact, has made it expressly mandatory that he shall exercise no internal security functions. Furthermore, this particular provision of the law, as the history of the legislation shows, was primarily intended to place upon the Director of Central Intelligence responsibility to see to the security of his own operations.

I have to admit, and do so with a mixture of regret and sadness, that during my years of service in the CIA I did not succeed in making much progress in this field. I did not find an acceptable and workable formula for tightening up our governmental machinery or slowing down the tempo of frustrating leaks of sensitive information of value to a potential enemy. For one must do this in the face of the understandable but sometimes uncontrolled yen of the press to know everything.

However, it should be possible to improve the situation, and I have felt that a frank discussion of the problem was in order. The British, through their Official Secrets Act and other related procedures, have a better legal system in this particular field than do we, and they are a country which prizes and protects the freedom of the press as do we. They have shown, however, that their practices in hiring and retaining personnel leave a good deal to be desired.

I start from the premise that nothing should be attempted which would affect the freedom of the press. Freedom, however, does not necessarily mean complete license where our national security is involved, and the First Amendment of the Constitution never intended this.

It will be difficult to try to deal with this phase of the problem of security through legislation, except in the tightening up of some of our espionage laws, as I shall explain. Rather, the government should put its own house in order by an understanding between the executive and the Congress and then seek the voluntary cooperation of the press.

Here is a possible order of procedure: (1) the executive branch of government, particularly the Departments of State and Defense and the intelligence community, should do what they can to prevent the unnecessary publication of information that is valuable to our enemies and to deal more effectively with the leaks from the executive branch; (2) in conference with Congressional leaders and in agreement with them, steps should be taken to restrict the publication of sensitive hearings in the field of our national security, particularly in the military field. After some progress has been made in (1) and (2), there should be quiet (hopefully) discussions between selected government officials most immediately concerned and the leaders of the press and other news media, radio, television, technical and service journals, to determine to what extent there can be mutual agreement for setting up machinery to keep the press confidentially advised as to the matters in which secrecy is essential to our security, particularly those pertaining to military hardware and sensitive intelligence operations.

Before doing this, it might well be worthwhile for the interested members of government and of the press to take a look at what has been accomplished in Great Britain through the D notice system, whereby on a voluntary basis the press cooperates with the government to prevent compromise of military secrets. In suggesting we study this system, I recognize that there are vital differences between the situation here and that in the British Isles, where there is such a large centralization of press and publications in one great city, namely, London. There is in this country no comparable center of authority in the matter of press and publicity, and it would be harder here to find any relatively restricted group of men in the field of news media whose judgment would be accepted by the press in all parts of the country. And in all fairness, I should also point out that the cooperation of the British press with the government is the result of the enforceability of the Official Secrets Act and is not in all cases purely voluntary. Newspapers frequently consult the government to be sure that material they intend to publish does not run counter to security standards. The D-notice system is over fifty years old, having been set up a year after the coming into force of the Official Secrets Act of 1911. It has no formal legal sanction but it operates through a committee consisting of four government representatives—the permanent heads of the War Office, the Admiralty, the Air Ministry and the Ministry of Aviation—and eleven representatives of the various news media. Where there is a sensitive national security matter which might well leak to the press, the secretary convenes the committee and the facts are presented. If all the press members concur, the notice goes out to the press. In urgent cases, the secretary is authorized to issue a D notice on his own responsibility but with the concurrence of at least two press members. If later other press members object to the D notice, it may have to be withdrawn, although this situation has never arisen, since the emergency powers have only been exercised on the rarest occasions where time was of the essence. The range of subjects covered by D notices are military matters, the publication of which would be prejudicial to the national interest, but the press does not insist on a rigid interpretation of this formula. A recent report of a committee headed by Lord Radcliffe, which was reviewing British security problems, also considered the effectiveness of the D notice system. It commented that "There have been cases of non-observance . . . more often accidental than deliberate and they have never been persisted in after the secretary has taken the matter up with the responsible editor." By its operation, the Radcliffe report indicates, the British government has succeeded "year in and year out in keeping out of newspapers, radio, and television a great deal of material . . . which needs to be concealed and which would be useful to other powers to possess . . . and which so far as we can see could not have been kept out in any other way." The Radcliffe report, in stressing that the D notice procedure "appears to suit the needs of both sides," added that according to the evidence before the committee "neither side wishes to amend the present system" and it recommended the continuance of the system along the present lines.

The point of studying this system would obviously be to see whether any of its features could usefully be adopted in this country to help deal with our own security problem. I would add that this procedure has nothing whatever to do with the case which has been much discussed on both sides of the Atlantic of the two British newsmen who spent several months in jail because they refused to tell a tribunal set up by Parliament to investigate the case of William Vassall the sources of stories they had written about him. There was a third reporter, who escaped a jail sentence because his reputed source voluntarily came forward and admitted to being the one who was the origin of the information. There are times, of course, when sources are not given because the writers would have some difficulty in producing them, even if they were so minded, as their stories might have been the product of their own intelligent guesswork. In the case of able reporters, these guesses often hit quite close to the mark.

A further point in the program to improve our security posture is that we should review and tighten up our espionage laws in certain respects. Since 1946, on several occasions, attempts, all abortive, have been made by the executive branch of government to amend the Espionage Act so that prosecution would not fail merely because of difficulties in establishing "an intent or reason to believe" that the information wrongly divulged or passed to a foreign government was "to be used to the injury of the United States or to the advantage of a foreign nation." This is hard to prove. Fortunately, the requirement of proof of such intent has already been eliminated in cases involving restricted data under the Atomic Energy Act and with regard to disclosure of classified information in the field of "communications intelligence." The requirement still holds, however, in cases where other types of secret and classified information are divulged. Much secret information has been divulged without authorization, even passed to foreign governments, where the defense would be made that the culprit was really trying to help our government by helping an ally—as the Soviet Union was for a time after 1941. There are other problems of a security nature which arise under our existing legislation when it is necessary to prove that a case is related to "the national defense and security," as our present espionage law requires.

Comparable British legislation is based on the theory of privilege, that all official information belongs to the Crown and that those who receive it officially may not lawfully divulge it without the authority of the Crown. This theory of government privilege in such matters seems a sound one. In our country, there are many cases where the disclosure in court of all the details of secret information wrongfully acquired or retained or passed on to the adversary may be contrary to the public interest. There are even times when prosecution has to be abandoned rather than divulge this classified information. Some persons who have been guilty of serious actions affecting our security were never prosecuted for one or more of the above reasons. The knowledge that our government is only likely to prosecute in the most heinous cases of espionage gives certain people the assurance that they can commit minor infringements against the espionage laws with impunity. The knowledge has not been lost on the Soviets.

If we drive a car in the streets with reckless abandon and inflict injury to life or property, there is no difficulty in prosecuting; but if our innermost secrets are handled with carelessness, there is little that can be done about it.

Even if we could plug up the holes in our espionage and security legislation—even if we could stop some of the giveaway of information of value to the enemy, there would still remain the dangers of human betrayal. By that I mean our own defectors and all those who betray our secrets and those of NATO, under alien pressure and blackmail, for money or for "ideological" reasons, or merely to satisfy their ego and exchange excitement for boredom. Here the watchful eye of government in a free society cannot provide adequate protective measures without appearing to infringe on the rights of the individual citizen. Unfortunately, there also are cases, here and abroad, when the eye of government has not been watchful enough. Too often the betrayer can act before the security services can catch up with him.

In addition to the prewar and wartime espionage cases, there have been Burgess, Maclean, and Philby; Houghton, Vassall and Blake in Britain, and more recently, Col. Wennerstrom in Sweden, Paques in France, and Dunlap here, who betrayed their trust. Also on our side the defection in 1960 of the two technicians from the National Security Agency, William H. Martin and Bernon F. Mitchell, was a shocker, and the betrayal by Irvin Scarbeck, the sordid affair of a weakling.

Perhaps the most disturbing treason case of all on our side of the water in recent years, from the point of view of the efficacy of our security practices, was that of Sergeant Jack E. Dunlap, who committed suicide on July 23, 1963, apparently because he could not face the consequences of the discovery and public exposure of his treasonable acts, which, though slow in coming, was inevitable. Dunlap, like Martin and Mitchell, was employed by the National Security Agency, but unlike them he did not occupy a position of any importance and did not have their specialized knowledge of highly sensitive communications matters. Instead, Dunlap's case was one frequently encountered in intelligence history where an insignificant employee, of meager understanding and less education, performing menial tasks but located at a vulnerable point in the internal workings of a highly secret undertaking, can do as much damage as a top-ranking official.

He was primarily a messenger and clerk responsible for the distribution and circulation of documents within NSA. What was in these documents may not even have been entirely intelligible to him. But it didn't have to be. All Dunlap had to do was photograph them and make sure that the film reached the Soviet officer handling him. If Dunlap had lived, it is unlikely that he could have recalled more than a small part of the material he passed the Soviets. The indication, however, that the documents were of value to the Soviets can be derived from the fact that Dunlap had very large sums of money at his disposal, owned fancy power boats, racing cars, had mistresses, etc. This affluent mode of living, highly suspicious in the case of a $100-a-week sergeant, did not come to the attention of his superiors for the simple reason that they could not and did not, under our present system, keep tabs on the private lives of their employees. Furthermore, as a member of the armed forces, Dunlap was not subject to the polygraph tests which are normally required of civilians in the NSA. Only when he left the Army and converted to civilian status did he have to submit to such a test, and it was on this occasion that the first suspicions were aroused concerning him. He was then removed from his position as a handler of sensitive documents because his reactions to the polygraph showed that he was not entirely trustworthy. This was the beginning of the end. Investigation and further polygraph tests were to follow. Dunlap obviously saw the handwriting on the wall. At least it can be said that the polygraph, as an indicator that something was amiss, which often is all that can be expected of it, did its work.

While the possible security implications of the Profumo–Ward–Christine Keeler–Ivanov "quadrangle" may never be ascertainable, we do know that here a Soviet intelligence officer, Yevgeni Ivanov, helped to undermine a government and its leaders. Thus he accomplished more to damage the Free World, whether by accident or design, than if he had obtained the intelligence information which he was apparently seeking.

In passing, it is worth noting that the exposure of presumed espionage or treason—indeed, even the hint of it—in high places has a powerfully disruptive effect on governments that can be matched by little else. The most notorious instance of this kind is, of course, the Dreyfus case, which rocked the French government and kept its political and military leaders embroiled and embattled for over a decade. It must often have occurred to the Soviets that if high-ranking members of a rival nation could be tainted with espionage, if only by implication, the advantages in the form of disruption, paralysis and dismay might far outweigh the rewards of successful espionage itself.

These and the other cases I have described earlier do show the inherent weaknesses of our free societies in protecting our nation's security. While there is a temptation here to point the finger at the security services, the real cause of the trouble is deeper.

The security services in England, and the same is largely true in the United States, generally have little to do with the security and personnel procedures and practices of other sensitive agencies of government. In the Profumo case, as far as I can judge, these security services had no basis for intervening until the Soviet agent Ivanov appeared on the scene; and with this, a possible breach of security loomed up. If before this the services had been caught out spying on the private lives of British subjects, not to speak of high government officials, then indeed there would have been an uproar.

In Britain, the foreign office and the defense agencies hire their own personnel, and often it is only when those so hired have already become security risks that the security services are called in. Then the damage has generally been done. Neither a Burgess nor a Maclean should ever have been allowed to have anything to do with classified matters. Even a reasonably casual review of their activities during the years before their defection should have resulted in their dismissal, and Burgess never should have been hired in the first place. In the case of Martin and Mitchell, I am convinced that if anyone had reported on the manner of their lives, an investigation would have resulted. Their living quarters were a shambles of disorder and slovenliness. Something must be wrong with people who lived as they did.

Under our system, and it is much the same in Britain, the security services do not continually go prying around into the private lives and private affairs of employees. We should have no Gestapo. A man's home is his castle, and it is sometimes suggested that a man's "private" life is of no concern as long as he does his work passably well.

Maybe the British, and maybe we, carry these principles too far. Government service, after all, is a privilege not a right, and to retain a government position one should live up to certain standards of moral conduct, standards which should be higher than those applied to others. The fact that one wears "the old school tie" is not enough.

In the Profumo case, it was emphasized in Parliament that security, not morals, was the main concern. Politically this may have been an astute line to take. The British press itself expressed editorial opinions to the general effect that one should not cast too many stones at the sexually wayward. One paper suggested, "On this basis, England would frequently have gone headless and guideless." The press pointed out that Nelson, to the distress of his wife, lived in flagrant and public adultery; that the Duke of Wellington was asked by Miss Harriette Wilson, described as the approximate equivalent in that day of Miss Christine Keeler, to pay her adequately for her agreement not to include an account of their relationship in her memoirs. "Publish and be damned" he replied. The British press pointed out that some of the most esteemed British leaders were not always models of moral propriety.

But these items of somewhat ancient British history related to men of courage in high positions, answerable for their conduct to the people as a whole. Also they occurred before we had the problem of Soviet intrigue and Soviet recruitment of the weak and wayward through blackmail. The conditions of past centuries are not a useful guide in personnel recruitment and job retention in the sensitive branches of government today. I see no reason why anyone should be employed or retained in a sensitive government position when there is credible evidence that the person has a serious character weakness or aberrations of conduct which might make him a possible victim for blackmail.

Of course, the problem of personnel security clearances becomes infinitely complicated because this requires periodic assessments and not just the one-time "vetting," which is the word the British use for it. People whose lives and records appear clean as a whistle when they are employed may, some years later, develop latent weaknesses, which may or may not be discovered in the course of security reviews. No one can suggest that even the most careful and the most frequent security examinations will point up all weaknesses. The best one can do is to have the most thorough examination that can be given, and I feel that one should not exclude, in the examination, technical aids such as the polygraph, more popularly known as the lie detector. In my experience, I found the "lie detector" an important investigative aid in sizing up employees and almost as valuable in clearing people of suspicious and false charges as it was in providing clues to weaknesses or derelictions.

It is dangerous to boast, but I can say that the security record of the CIA has been an excellent one. It got off to a good start and was greatly strengthened under that stern but understanding disciplinarian, General Walter Bedell Smith, my predecessor in CIA, who worked out the principles of its firm security practices. "Beedle" once shocked the public and the press by stating in 1952—this was during the electoral campaign of that fall—that one must assume that there could be a Soviet agent in the CIA. He was quite right to sound such an alarm. One must always assume that there is a chance that this is the case, though neither of us was able to find such a culprit. With the years that have elapsed, we may be more optimistic but never sure as to whether or not he is there. The fact that we had reasonable success on the security side in the CIA was not because of any complacency or failure to try to ferret out the facts or any tendency to "cover up" as is so often done. We set out to eliminate from consideration as employees all known homosexuals, persons of unstable or weak character, or persons whose home or family life seemed likely to produce instability. In this, the Agency has had reasonable success.

In our own government setup, there is a security office in each sensitive agency which has responsibility for the security of that particular agency. The Civil Service Commission and, on occasions, the FBI assist in the investigations of employees of other agencies. Their task is generally limited to checking out an individual by means of interviews with associates, neighbors and others who could cast some light on the prospective employee's character. They will also check other government records. They do not decide whether a person should or should not be employed. Final responsibility for personnel security decisions rests with the particular agency doing the hiring or the firing.

Each security office, in State, Defense and the Armed Services, the National Security Agency and the Atomic Energy Commission, as well as in CIA, should profit by the experiences of the others, and there is, of course, coordination and consultation among them. As various methods are tried out to eliminate security risks, experiences are exchanged. In some departments, the importance of continuity of service and of experience has not been adequately stressed in choosing the head of the division of security. The idea that this job is one which a political appointee may hold down for a year or two is dangerous. This is a task for the trained professional who should look to a long period of service.

On the other hand, I feel that our industrial leaders for the most part have taken very seriously the protection of the security of their plants where classified work for the government is being carried on. The large number of workers in many of these plants which have various secret phases of our military production in hand makes this a difficult task. I had a striking example of the effectiveness of industrial security during my work on the U-2 and later during the initial stages of the A-11, the first and second generation of the renowned reconnaissance aircraft. Despite the dramatic innovations introduced by these aircraft and the particularly high sensitivity of this work, the Lockheed Company, the builder of the planes, maintained throughout a high degree of security.

I would add a word about the security of our overseas installations where sensitive work is carried on. These are chiefly our various embassies throughout the world and, in certain places where we have overseas forces, sensitive military installations. As compared with the Soviet, it would appear that we are rather lax. Their overseas missions, particularly their embassies, are made as far as possible into self-contained fortresses. Except for social occasions, few outsiders are admitted. As far as possible, they have their own personnel to take care of even minor housekeeping problems, such as plumbing, electricity, minor repairs and the like. They rarely, if ever, employ outside local personnel or give them free access to their installations.

The Soviet embassy premises in Teheran made a great impression on me when I saw them a few years ago. They took up a whole city block, were surrounded entirely by high walls, guarded at all vulnerable points—in short, a real fortress. And the Soviets like to house as many as possible of their own personnel within their embassies so that they can keep good watch on them.

I would not emulate all such security precautions of the Soviets, and we have no need to turn our embassies into fortresses or to segregate within embassy premises all of our personnel. But in many instances behind the Iron Curtain we make too much use of local personnel, something the Soviets would never think of doing. This was pointed up in the report made to Parliament in 1963 by the tribunal appointed under the Inquiry Act of 1921 to look into the Vassall case. This tribunal was also headed by Lord Radcliffe, whose earlier report on the Lonsdale affair I have mentioned above.[5]

[5] The two Radcliffe reports referred to in this chapter are of interest to those who deal with security matters. The first, published in a Parliamentary paper in April 1962 (cmnd. 1681), followed the Lonsdale spy ring case (the British call it the Portland case) and the Blake case. It proposed some tightening up of British security practices and also dealt with the problem of security and press relations. The second Radcliffe report, published in April 1963 (cmnd. 2009), was that of a tribunal established under the Tribunals of Inquiry Act of 1921 and a Parliament resolution of November 14, 1962. Viscount Radcliffe was the Chairman of the earlier committee and of the tribunal. The tribunal's report was limited to a judicial investigation of the circumstances of the Vassall case. It is an interesting contribution but, being a legal judgment, it naturally tends to stress the question as to whether there were or were not clear-cut derelictions of duty. Hence it is not so useful in trying to determine the wisdom of the judgments reached by Vassall's superiors in the years prior to his arrest in the light of all our present knowledge of the sinister character of Soviet penetration techniques.

The British embassy in Moscow during the days that Vassall was in the Naval Attaché's office there employed a factotum named Mikhailski, a Pole described in the Radcliffe report as an "agent of the Russian secret service who was the instrument by which they secured their hold on Vassall." He acted, the report states, "as an assistant in the administrative section of the Embassy" and "made himself useful to the Embassy staff as an interpreter and local agent for arranging such matters as helping with Russian servants, travel facilities" and the like. In this capacity, he was "of real importance in contributing to the ease and convenience of the British staff, particularly with the language difficulty between English and Russian, and somehow they must be provided for in the general interests of staff morale." The Radcliffe report recognized that this constituted "a fixed security risk" and so it proved to be in the Vassall case. While the Radcliffe report tends to exonerate the employment of the man because of the great convenience it represented, I must say that I think such a practice behind the Iron Curtain is dangerous and one to be discouraged. Certainly security should take precedence over convenience, and we would do better in these countries to man all our sensitive missions, diplomatic and military, with American personnel from top to bottom.

Actually the United States and Britain are not alone in requiring and using the services of local personnel in Moscow. Each foreign country, including the Soviet satellites, which maintains an embassy there—with the notable exception of the Communist Chinese—hires native help for jobs such as chauffeurs, cleaners, purchasing agents and the like. By bringing all their own personnel along down to the lowliest charwoman, the Chinese in Moscow enjoy the same improved security that the Soviets maintain in all their own installations abroad.

The business of providing sufficient housekeeping personnel to meet the needs of the many foreign embassies in Moscow is such a large one that the Soviet government has a special bureau, a kind of employment office, called BUROBIN, which supplies the needed help on request. This is obviously a highly organized KGB-controlled clearing house for agents who are trained to make the most of their jobs in foreign installations. BUROBIN will assign English-speaking chauffeurs or cleaning women on its roster to the British or Americans when the latter ask for personnel, French-speakers to the French, and so on. A chauffeur of the American Embassy, a Soviet national, incidentally played a mysterious role in the frame-up of Professor Barghoorn in Moscow in the fall of 1963.

The fact that in recent times the Western world has turned up a large number of Soviet espionage operations should not necessarily lead us to the conclusion that our security services are ineffective. On the contrary, it is the best evidence we could have that our counterintelligence, which is the offensive arm of our security, is strong. Thanks to it we are now uncovering Soviet penetrations that have gone undetected for many years. Although some embarrassment on our side is unavoidable, the Soviets are the ones who have received the rudest shock, and they may be forced as a result to overhaul many of their espionage techniques. At the same time, these belated discoveries of Soviet agents in our midst should serve as a warning to us of the depth and sophistication of the Kremlin's espionage effort and should make us more understanding of the need to tighten our own security practices in order to prevent the possibility of such penetrations in the first place.

16

The Intelligence Service and Our Freedoms

From time to time the charge is made that an intelligence or security service may become a threat to our own freedoms, that the secrecy under which such a service must necessarily operate is in itself vaguely sinister and that its activities may be inconsistent with the principles of a free society. There has been some sensational writing about the CIA's supposedly supporting dictators, making national policy on its own and playing fast and loose with its secret funds. Harry Howe Ransom, who has written a study on *Central Intelligence and National Security*, puts the issue this way:

> CIA is the indispensable gatherer and evaluator of worldwide facts for the National Security Council. Yet to most persons CIA remains a mysterious, super-secret shadow agency of government. Its invisible role, its power and influence, and the secrecy enshrouding its structure and operations raise important questions regarding its place in the democratic process. One such question is: How shall a democracy insure that its secret intelligence apparatus becomes neither a vehicle for conspiracy nor a suppressor of the traditional liberties of democratic self-government?[1]

1 *Central Intelligence and National Security* (Cambridge, Mass: Harvard University Press, 1958).

It is understandable that a relatively new organization in our government's structure like the CIA should, despite its desire for anonymity, receive more than its share of publicity and be subject to questioning and to attack. In writing this analysis of intelligence, I have been motivated by the desire to put intelligence in our free society in its proper perspective. As I have already indicated, CIA is a publicly recognized institution of government. Its duties, its place in the official hierarchy and the controls surrounding it are set forth partly by statutes, partly by National Security Council directives. At the same time, as is true for other departments of government, some of its work must be kept secret.

I have already pointed out that in both Czarist and Soviet Russia, in Germany under Hitler, in Japan under the war lords and in certain other countries where dictators ruled, security services that exercised some intelligence functions were used to help a tyrant or a totalitarian society suppress freedoms at home and carry out terrorist operations both at home and abroad.

Moreover, as I have already suggested, there have been many instances—most conspicuously in Latin America—in which dictators have converted authentic intelligence services into private gestapos for maintaining their rule.

This warped use of the intelligence apparatus and the wide notoriety it has obtained have tended to confuse many people about the true functions of an intelligence service in a free society.

Our government in its very nature—and our open society in all its instincts—under the Constitution and the Bill of Rights automatically outlaws intelligence organizations of the kind that have developed in police states. Such organizations as Himmler's Gestapo and the Kremlin's KGB could never take root in this country. The law which set up CIA specifically provides "that the Agency shall have no police, subpoena, law-enforcement powers, or internal security functions." Furthermore, it is the servant, not the maker, of policy. All its actions must stem from and accord with settled national policy. It cannot act without the authority and approval of the highest policymaking organizations of the government.

The legislation, which was adopted with bipartisan support, also threw other legal and practical safeguards around the work of the CIA. These accorded for the most part with the safeguards that protect any democracy.

The Central Intelligence Agency is placed directly under the National Security Council and is, therefore, immediately under the President. Thus it is the Chief Executive himself who has the final responsibility for overseeing the operations of the CIA.

The National Security Council directives are issued under the authority of the National Security Act of 1947, which provides that, in addition to the duties and functions specifically assigned under law, the CIA is further empowered to

> perform for the benefit of the existing intelligence agencies such additional services of common concern as the National Security Council determines can be more efficiently accomplished centrally . . . and perform such other functions and duties relating to intelligence affecting the national security as the National Security Council may from time to time direct.

It is the President who selects, and the Senate which confirms, the Director and the Deputy Director of the Agency, and this choice is no routine affair. In the fifteen years since the Agency was created, it has had four Directors: (1) Rear Admiral Roscoe Henry Hillenkoetter, who had distinguished service in the Navy and in Naval Intelligence; (2) General Walter Bedell Smith, who, in addition to an outstanding military career, for almost three years was American Ambassador to the Soviet Union before he was Director and, afterward, Under Secretary of State; (3) the writer—and here any comment by me would be out of place, except at least to mention a long period of government service and many years in intelligence work; and (4) John A. McCone, who before being named Director in 1961 had done outstanding service in both the Truman and the Eisenhower administration in many important government posts—as a member of the President's Air Policy Commission, as a Deputy to the Secretary of Defense, as Under Secretary of the Air Force, and then as Chairman of the United States Atomic Energy Commission.

The law provides that a civilian must be in the position of either Director or Deputy Director. While, theoretically, it is possible to have both of these jobs in civilian hands, military men cannot fill both positions as the law now stands. (The practice over the past decade has been to split them between a military man and a civilian.) The last two Directors, both civilians, have had highly experienced military men for Deputy

Directors—General Charles Pearre Cabell during my tenure, and Lieutenant General Marshall S. Carter under John McCone.

From my own experience in the Agency, under three Presidents, I can say with certainty that the Chief Executive takes a deep and continuing interest in its operations. During eight of my eleven years as Deputy Director and Director of the CIA, I served under President Eisenhower. I had many talks with him about the day-to-day workings of the Agency, particularly concerning the handling of its funds. I recall his instructing me that we should set up procedures in the Agency for the internal accounting of unvouchered funds, i.e., funds appropriated by Congress and expendable on the signature of the Director, which would be even more searching, if that were possible, than those of the General Accounting Office. This was done.

While obviously many expenditures must be kept secret as far as the public is concerned, the CIA always stands ready to account to the President, to the responsible appropriations subcommittees of the Congress and to the Bureau of the Budget for every penny expended, whatever the purpose.

During the earlier years of the Agency, there was a series of special investigations of its activities. I myself, as I have mentioned, was the head of a committee of three that in 1949 reported to President Truman on CIA operations. There were also studies made under the auspices of two Hoover Commissions, one in 1949 and one in 1955. These dealt with the organization of the executive branch of government and included studies on our intelligence structure. The survey conducted in 1955, during my directorship, included a report prepared by a task force under the leadership of General Mark W. Clark; at about the same time, a special survey of certain of the more secret operations of the Agency was prepared for President Eisenhower by a task force under General James Doolittle. It is interesting to note that General Clark's task force, expressing concern over the dearth of intelligence data from behind the Iron Curtain, called for "aggressive leadership, boldness and persistence." We were urged to do more, not less—the U-2 was already on the drawing boards and was to fly within the year.

One of the recommendations that emerged from the Hoover Commission survey in 1955 called for establishing a permanent Presidential civilian board, often called a watchdog committee. This would take the place of ad hoc investigative committees. I discussed with President Eisenhower how this could best be done. Personally, I strongly favored the idea. He appointed a "President's Board of Consultants on Foreign Intelligence Activities," which for some time was chaired by the distinguished head of the Massachusetts Institute of Technology, James R. Killian, Jr. President Kennedy, shortly after he took office, reconstituted this Presidential committee with a slightly modified membership and again under the chairmanship of Dr. Killian. In April, 1963, Dr. Killian resigned, and an eminent lawyer and expert in government, Clark Clifford, succeeded him as chairman. The files, the records, the activities, the expenditures of the Central Intelligence Agency are open to this Presidential committee, which meets several times a year.

The other recommendation of the Hoover Commission, that a Congressional watchdog committee should also be considered, had a somewhat more stormy history.

In 1953, even before the Hoover recommendations, Senator Mike Mansfield had introduced a bill to establish a joint Congressional committee for the CIA, somewhat along the lines of the Joint Committee on Atomic Energy. On August 25, 1953, he wrote me a letter to inquire about CIA's relations with Congress and asked the Agency's views on the resolution he had submitted. In my absence abroad, General Cabell, my deputy, replied that "the ties of the CIA with the Congress are stronger than those which exist between any other nation's intelligence service and its legislative body." In fact, I can state with assurance that CIA is today the most "watched over" intelligence organization in the world.

A few years later this issue came to a vote in the Senate in the form of a concurrent resolution sponsored by Senator Mansfield. It had considerable support, as thirty-five Senators from both parties were cosponsors, and the resolution had been reported out favorably by the Senate Rules Committee in February of 1956, but one vote of strong dissent came from Senator Carl Hayden, who was also the chairman of the Senate Appropriations Committee. Senator Hayden's viewpoint was supported by Senator Richard Russell, chairman of the Senate Armed Services Committee, and by Senator Leverett Saltonstall, the senior Republican member of that committee. In April, after a most interesting debate, the Senate voted against the watchdog committee resolution by a surprisingly large majority. In opposing the resolution, Senator Russell said: "Although we have asked him [Allen W. Dulles] very searching questions about some activities which it almost chills the marrow of a man to hear about, he has never failed to answer us forthrightly and frankly in response to any questions we have asked him." The issue was decided when this testimony was supported by former Vice President (then Senator) Alben Barkley, who spoke from his experience as a member of the National Security Council. He was joined in opposition by Senator Stuart Symington, who had intimate knowledge of the workings of the Agency from his days as Secretary of the Air Force. On the final vote of 59 to 27, ten of the measure's original cosponsors reversed their positions and joined with the majority to defeat the proposal. They had heard enough to persuade them that the measure was not needed.

During the debate it was pointed out with a great deal of emphasis that procedures serving the intended end had already been set up and had been functioning well for some years.

Any public impression that the Congress exerts no power over CIA is quite mistaken. Control of funds gives a control over the scope of operations—how many people CIA can employ, how much it can do and to some extent what it can't do. Even before a Congressional subcommittee sees the CIA budget, there is a review by the Bureau of the Budget, which must approve the amount set aside for CIA, and this, of course, includes Presidential approval. Then the budget is considered by subcommittees of the Appropriations Committees of the House and of the Senate, as is the case with other executive departments and agencies. The only difference in the case of the CIA is that the amount of its budget is not publicly disclosed, except to these subcommittees.

The chairman of the House subcommittee for many years and until his death in 1964 was Clarence Cannon, and a more careful watchdog of the public treasury could hardly be found. This subcommittee is entitled to see everything it wishes to see with regard to the CIA budget and to have as much explanation of expenditures, past and present, as it desires.

All this was clearly brought out in a dramatic statement that Mr. Cannon made on the floor of the House on May 10, 1960, just after the failure of the U-2 flight of Francis Gary Powers: "The plane was on an espionage mission authorized and supported by money provided under an appropriation recommended by the House Committee on Appropriations and passed by the Congress."

He then referred to the fact that the appropriation and the activity had also been approved and recommended by the Bureau of the Budget and, like all such expenditures and operations, was under the aegis of the Chief Executive. He discussed the authority of the subcommittee of the Appropriations Committee to recommend an appropriation for such purposes and also the fact that these activities had not been divulged to the House and to the country. He recalled the circumstances during World War II when billions of dollars were appropriated, through the Manhattan Project, for the atomic bomb under the same general safeguards as in the case of the U-2, i.e., on the authority of a subcommittee of the Appropriations Committee. He referred to the widespread espionage by the Soviet Union, to the activities of their spies in stealing the secret of the atomic bomb. Alluding to the surprise attack by the Communists in Korea in 1950, he justified the U-2 operation in these words:

> Each year we have admonished . . . the CIA that it must meet situations of this character with effective measures. We told them, "This must not happen again and it is up to you to see that it does not happen again" . . . and the plan that they were following when the plane was taken is their answer to that demand.

Mr. Cannon took occasion to commend the CIA for its action in sending reconnaissance planes over the Soviet Union for the four years preceding Powers' capture and concluded:

> We have here demonstrated conclusively that free men, confronted by the most ruthless and criminal despotism, can under the Constitution of the United States protect this nation and preserve world civilization.

I cite this merely to show the extent to which even the most secret of the CIA's intelligence operations have, under appropriate safeguards, been laid before the representatives of the people in Congress.

In addition to the scrutiny of CIA activities by the Appropriations Committee, there is also a subcommittee of the House Armed Services Committee, chaired by Congressman Carl Vinson, who for years has been head of the Armed Services Committee itself. To this subcommittee, the Agency reports its current operations to the extent and in the detail the committee desires, dealing here not so much with the financial aspects of operations but with all the other elements of CIA's work. In the Senate, there is a comparable subcommittee of the Armed Services Committee.

Fifteen years ago, when the legislation to set up a Central Intelligence Agency was being considered, the Congressional committees working on the matter sought my views. In addition to testifying, I submitted a memorandum, published in the record of the proceedings, in which I proposed that a special advisory body for the new Agency should be constituted to include representatives of the President, the Secretary of State and the Secretary of Defense. This group should, I proposed, "assume the responsibility for advising and counseling the Director of Intelligence and assure the proper liaison between the Agency and these two Departments and the Executive." This procedure has been followed. All operations of an intelligence character which involve policy considerations are subject to such approval.

Of course, the public and the press remain free to criticize the actions taken by intelligence, including those which are exposed by mishap or indiscretion. This holds just as true for intelligence activities as for any government operations. When an intelligence operation goes wrong and publicity results, the Intelligence Agency and its Director, in particular, must stand ready to assume responsibility if silence is impossible. There have been times, as in the case of the U-2 descent on Soviet territory, and the Cuban affair of April, 1961, where the executive has publicly assumed responsibility, and for good reasons, as I have already explained.

It is an established rule that the Agency should never intervene in policy matters except when and where specifically directed by high authority.

Also, its personnel should keep out of politics. No one in the Agency, from the Director on down, may engage in any political activities of any nature, except to vote. A resignation is immediately accepted—or demanded—where this occurs, and the political aspirant is given to understand that quick reemployment, in case of any unsuccessful plunge into the political arena, is unlikely.

In the last analysis, however, the most important safeguards lie in the character and self-discipline of the leadership of the intelligence service and of the people who work for it—on the kind of men and women we have on the job, their integrity and their respect for the democratic processes and their sense of duty and devotion in carrying out their important and delicate tasks.

After more than a decade of service, I can testify that I have never known a group of men and women more devoted in the defense of our country and its way of life than those who are working in the Central Intelligence Agency. Our people do not go into intelligence for financial reward or because the service can give them, in return for their work, high rank or public acclaim. They do it because of the opportunity to serve their country, the fascination of the work and the belief that through this service they personally can make a contribution to our nation's security.

It is not our intelligence organization which threatens our liberties. The threat is rather that we will not be adequately informed of the perils which face us and that we will fail to act in time. If we have more Cubas, if non-Communist countries which are today in jeopardy are further weakened, then we could well be isolated and our own liberties, too, could be in danger.

The military challenge of the nuclear missile age is well understood, and we are rightly spending billions to counter it. We must also deal with all aspects of the invisible war, the Kremlin's war of liberation, the subversive threats orchestrated by the Soviet Communist party with all its ramifications and fronts, supported by espionage. The last thing we can afford to do today is to put our intelligence in chains. Its protective and informative role is indispensable in an era of unique and continuing danger.

Bibliography

The following is a selected list of a few texts on various aspects of intelligence, available in English, which I have found useful in the preparation of this book on "The Craft of Intelligence."

1. HISTORICAL

Richard Wilmer Rowan. *The Story of Secret Service.* (Garden City, N.Y.: Doubleday. 1937.) A comprehensive history of espionage and its practitioners from Bible days to the end of World War I.

Samuel B. Griffith. *Sun Tzu, The Art of War.* (Oxford: Clarendon Press. 1963.) A modern translation of, and commentary on, this ancient and important Chinese work.

John Bakeless. *Turncoats, Traitors and Heroes.* (Philadelphia: J. B. Lippincott. 1959.) An account of espionage in the American Revolution.

Barbara W. Tuchman. *The Zimmermann Telegram.* (New York: Viking. 1958.) A recounting of the most significant achievement in cryptanalysis during World War I.

Ewen Edward Samuel Montagu. *The Man Who Never Was.* (Philadelphia: J. B. Lippincott. 1954.) A reconstruction, by the man who was in charge of the operation, of the classic British hoax in World War II which misled the Nazis about the coming Allied invasion of Sicily.

Stewart Alsop and Thomas Braden. *Sub Rosa; the O.S.S. and American Espionage.* (New York: Reynal and Hitchcock. 1946.) Examples of OSS clandestine intelligence and paramilitary operations in Europe, Africa and Asia.

2. THE U. S. GOVERNMENT AND INTELLIGENCE

Sherman Kent. *Strategic Intelligence for American World Policy.* (Princeton, N.J.: Princeton University Press. 1949.) The theory and operation of national intelligence production.

Harry Howe Ransom. *Central Intelligence and National Security.* (Cambridge, Mass.: Harvard University Press. 1958.) Development, organization and problems of the U.S. intelligence system.

Douglass Cater. *The Fourth Branch of Government.* (Boston: Houghton Mifflin. 1959.) The government, the press and security.

3. SOVIET INTELLIGENCE

Simon Wolin and Robert M. Slusser. *The Soviet Secret Police.* (New York: Frederick A. Praeger. 1957.) History of the Soviet State Security services from the Cheka in 1917 until 1956.

David J. Dallin. *Soviet Espionage.* (New Haven: Yale University Press. 1955.) A thorough account of Soviet foreign intelligence from the 1920s until 1954.

Alexander Foote. *Handbook for Spies.* (Garden City, N.Y.: Doubleday. 1949.) The case history of the operation of a Soviet wartime intelligence net.

Alan Moorehead. *The Traitors.* (New York: Harper & Row. 1963.) Atomic espionage of the World War II period, with particular emphasis on Klaus Fuchs.

John Bullock and Henry Miller. *Spy Ring: A Story of the Naval Secrets Case.* (London: Secker and Warburg. 1961.) A blow-by-blow description of the activities and eventual apprehension of the Soviet net in England headed by Gordon Lonsdale.

4. BOOKS GIVING THE EXPERIENCES OF FORMER OFFICERS
OF SOVIET BLOC INTELLIGENCE SERVICES

Peter Deriabin and Frank Gibney, *The Secret World*. (Garden City, N.Y.: Doubleday. 1959.) A presentation of the organization and functions of Soviet State Security from 1946 to 1953.

Igor Gouzenko. *The Iron Curtain*. (New York: E. P. Dutton & Co. 1948.) The Soviet code clerk's own exposé of Soviet intelligence activities in Canada during World War II.

Vladimir and Evdokia Petrov. *Empire of Fear*. (New York: Frederick A. Praeger. 1957.) The Petrovs' own story of their intelligence activities in Australia on behalf of the U.S.S.R.

Pawel Monat. *Spy in the U.S.* (New York: Harper & Row. 1961.) Reminiscences and reflections of the former Polish Military Attaché in Washington on his own intelligence-gathering activities in the U.S.A.

Alexander Orlov. *Handbook of Intelligence and Guerrilla Warfare*. (Ann Arbor, Mich.: University of Michigan Press. 1963.) Orlov was one of Stalin's NKVD chiefs in Spain during the Civil War, at which time he defected. This is a book on the techniques of clandestine intelligence and clandestine warfare as practiced by the Soviets in the '20s and '30s.

Aleksandr Kaznacheev. *Inside a Soviet Embassy*. (Philadelphia: J. B. Lippincott. 1962.) Experiences of a Russian diplomat in Burma who was at the same time a secret member of the Soviet intelligence service.

5. THE COLD WAR

Josef Korbel. *The Communist Subversion of Czechoslovakia, 1938–1948*. (Princeton, New Jersey: Princeton University Press. 1959.) A report on the Communists' take-over of Czechoslovakia.

Dan Kurzman. *Subversion of the Innocents*. (New York: Random House. 1963.) A country-by-country account of Soviet attempts to infiltrate and take over weaker nations, uncommitted or otherwise, in Africa, the Middle East and Asia.

INDEX